Out of the Field:

Preserve Your Home

A Collection of Booklets
First Published by
International Harvester
Service Bureau

For my Emily, Jake and Cody. Love Always

Dear Reader,

I was asked recently, "Can you dehydrate Pumpkins?" Of course you can, was my reply. The discussion then led on to How, Why and how come one cannot find this information without some serious searching. With my passion to bring heritage and historic varieties of pumpkins, or winter squash, into the modern world, I have also noticed that not very many people actually preserve foods or simply budget their food for the week or month. I feel like I am constantly answering the question of "how long will this pumpkin last?" Winter Squashes and pumpkins are meant for long term storage, and generally are better eating when they 'sit' for a while in storage. Usually they are good for 3-4 months when stored properly and free of blemishes, however, I have had a pumpkin in storage for over a year!

One cannot always guarantee these will last without disastrous results, or spoilage, and so we turn to the preservation of food. By learning how to preserve our food, we can eat in the middle of winter and feel like summer is right around the corner- when opening a can of peaches or making a pie from your own preserved filling. Please DO NOT attempt to preserve any food by following these methods, the Ball Company has a very nice updated booklet that I would recommend using instead.

Now you are asking -

Why a Collection of booklets from one hundred years ago? You Just said not to use it as a guide! In 1912, International Harvester started the Service Bureau, an extension service that covered everything to do with farming, before, during and after planting, as well as help keep the farmer and the home healthy and educated. I cannot tell you how many times I have read the caption "The Home is producing the future men and women - the greatest crop of all." Keep this in mind while reading and you will understand the great love, importance and influence that Mrs. Cyrus McCormick (Nettie) had with her children and her husband. I feel throughout my research, she wanted the best future for her family, but did not need the recognition or fame, as long as her family was happy, so was she. Her family also happened to be just about everyone that had a farm or visited her home. She was very well-liked and an astute businesswoman, credited with encouraging Cyrus to rebuild after the Great Chicago Fire burned the factory down. She didn't need to rebuild, but the people the McCormick's employed needed her to, and they were family to her.

The Service Bureau was the precursor to the home goods line that was introduced after World War II; Irma Harding was the fictional Spokesperson, and the army of Harvester men and women who created her had their beginnings in the Bureau.

Many times throughout the booklets it is referred to the Wife as the domestic person. While Harvester knew that demographics throughout the country varied, they used the female as the inside worker, and the male as the outside worker, for simplicity. Quite often, they knew in order to sell more equipment, it had to benefit the whole family and the wife had the final say!

These Booklets brought me great joy reading them, to change them in any way would lose the authentic time period that they were written in. Knowing that there is a great loyalty to the best company in history, the Harvester Spirit is alive and well, always striving for better, adapting to the circumstances surrounding the farm and home, and making life in general better for the community we live in. Please remember that some ingredients have changed names or have since become unusable, and are unsafe to use under any circumstances suggested.

Happy Reading, Sarah

CONTENTS

PAGE

11   BOOK 1   A GOOD HOME PROVIDES COMFORT, PROFIT AND PLEASURE

35   BOOK 2   BUDGET YOUR FOOD SUPPLY

49   BOOK 3   GROW A VEGETABLE GARDEN

97   BOOK 4   HOME DRYING OF FRUITS AND VEGETABLES

117   BOOK 5   HOME CANNING BY THE COLD PACK METHOD (LECTURE SLIDE VERSION)

183   BOOK 6   AN ACRE HOME FOR EVERY AMERICAN

207   BOOK 7   HELPS FOR WASH DAY

THE FOLLOWING IS FOR HISTORICAL PURPOSES ONLY
AND DOES NOT REFLECT CURRENT
SCIENTIFIC KNOWLEDGE, POLICIES,
PRACTICES, METHODS OR USES.

THE FOLLOWING IS FOR HISTORICAL PURPOSES ONLY
AND DOES NOT REFLECT CURRENT
SCIENTIFIC KNOWLEDGE, POLICIES,
PRACTICES, METHODS OR USES.

A GOOD HOME

PROVIDES

COMFORT
PROFIT
PLEASURE

PUBLISHED 1917 BY
INTERNATIONAL HARVESTER COMPANY
OF NEW JERSEY (INCORPORATED)
AGRICULTURAL EXTENSION DEPARTMENT
HARVESTER BUILDING, CHICAGO

THE HOME

The home—that institution for which and by which all other institutions in the world exist.

Put the same intelligence and training into the making of the home that is given to great business enterprises.

The home is producing the future men and women—the greatest crop of all.

A
GOOD HOME

PROVIDES

COMFORT
PROFIT
PLEASURE

PUBLISHED 1917 BY
INTERNATIONAL HARVESTER COMPANY
OF NEW JERSEY (INCORPORATED)
AGRICULTURAL EXTENSION DEPARTMENT
P. G. HOLDEN, Director
HARVESTER BLDG., CHICAGO

AE 388 5-15-17

A Practical Power House for the Country Home, Where Power is
Generated for Pumping Water, Furnishing Light, Turning
the Washing Machine, Churn, Cream Separator,
Vacuum Cleaner, Etc. (See page 18.)

MAKE THE PLACE HOMELIKE

It's Not Homelike Without Trees, Flowers, Pictures, and a Garden

Make the Place Homelike—Make it convenient, sanitary; supply pure running water; equip with mechanical helps.

What Is a Home?—Is it a space enclosed between four walls, where we go at the end of the day's work for shelter? Or is it a place of charity and friendship, where life is lived at its fullest?

It was Irving Bacheller who so truly said in his story of Silas Strong:

"When you enter a home you begin to feel the heart of the owner. Something in the walls and furnishings; something in the air * * * that bids you welcome, or warns you to depart—it is the true voice of the master."

Make the Home Beautiful—A good home does not necessarily mean a mansion or a palace, but a simple structure made comfortable by location, arrangement, ventilation, sanitation, decoration and convenience.

A well-planned house, surrounded by trees, shrubs and grass, furnished for comfort rather than "show;" equipped with heat, running water, a bathroom, and labor-saving devices makes life anywhere a joy. The modern conveniences have been so perfected that they can be easily installed in any home and make possible all the comforts of the city in the country home.

Exterior Plan of Buildings and Grounds

A Good Home:

Location with slope away from house.
Color schemes: In harmony with Nature and surroundings. Plants and shrubs: Cover unsightly buildings and bare porches with quick-growing vines and woody twiners.
Plant shrubs and flowers to conceal angles and foundations, and to help make a natural picture of the premises. Plant trees for shade, for windbreaks, or for a background. Leave a smooth open lawn in front.
Make the approach direct, but not prominent..

Interior

Walls: Treated according to material, use and location of room.

A few well-framed pictures, some good vases.

Floors: Kitchen—Smooth, stained and oiled; linoleum.

Living Rooms—Stained and oiled, rugs. Beautiful rugs can be made at home of braided, woven or crocheted rags.

Windows: Curtains—Simple muslin.

Shades—Material to harmonize with walls. Dark shades make room look larger.

Woodwork: Plain; easily kept clean. Treated to suit use and location of room. In harmony with walls and furniture.

Furniture: Simple in outline, suitable to use. Easily kept clean. Have a few pretty house plants.

SAVING STEPS

Saving Steps Means Saving Time, Strength, Energy and Health—Means Comfort, Enjoyment and Success—Lifts Housework Above Drudgery

Is it not surprising to find that a saving housewife is often wasteful and extravagant? If we were to say: "The average housewife throws away $300 a year," or if we said: "She wastes a pound of sugar a week," we should begin at once to search for the reason and the remedy for the waste. Yet the waste of vital energy is much more extravagant and the consequences much more serious.

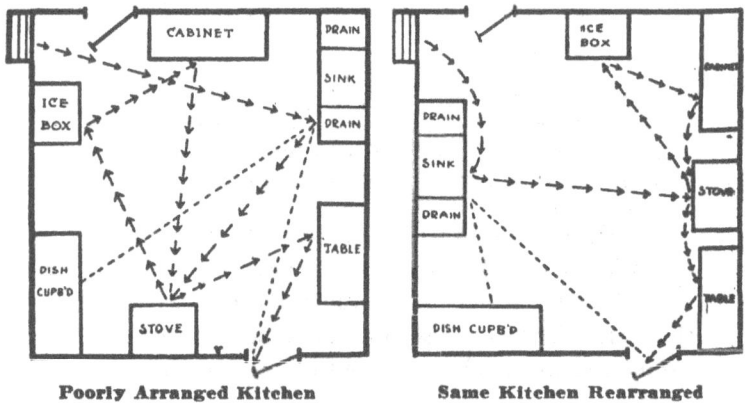

Poorly Arranged Kitchen   Same Kitchen Rearranged

The average woman wastes a vast amount of energy in useless walking. Her kitchen furnishings are arranged around the wall in such a way that she crosses and recrosses the central floor space from one place to another. Don't cross your tracks. Figure out for yourself the order in which you use your kitchen furniture and then arrange it in that order.

In most kitchens it will go something like this:

Materials from the cellar, icebox, cooler or cupboard are brought first to the sink for cleaning, then to the stove for cooking, from the stove they are dished up on the table, and then into the dining-room. Then why not work from left to right straight from storage place to serving?

In planning the kitchen we find there are just two main processes in all kitchen work. One process is to prepare the meal, the other to clear it away. Disorganization in the kitchen means wasted energy.

Standardize the arrangement of your kitchen by placing the equipment, stove, sink, icebox, cabinet or cupboard, table, etc., in right relation to each other. Material and furniture for clearing away should be aranged with the same object in view.

We have taken for example a kitchen measuring 15 feet by 15 feet in size. We measure the distance traveled in preparing one vegetable, from cellar to sink for the washing, to stove for cooking, to icebox or cooler and cabinet for butter or cream and seasoning, back to stove, to table for dishing up, to dining-room. The distance traveled was 75½ feet.

In clearing away this one dish the distance traveled was 32½ feet. In the second diagram we have shown the same kitchen with furnishings rearranged with a view to placing the equipment used for preparing the meal and that for clearing away in the right relation to each other. The distance traveled in preparing one vegetable with this arrangement was only 52½ feet, and in clearing away only 20 feet.

Figuring on this basis that we should prepare on an average five dishes for each meal and three meals a day for one month we should save more than 15,000 feet of walking by having the equipment arranged as in the second diagram. In one year the distance saved in steps would amount to more than 35 miles.

When the furniture has been arranged that the work may be done with the fewest steps, then study each task and have the right equipment and tools for each.

Is the dishpan below or within reach of the sink, or is it in a cupboard across the room? Is there a little shelf over the sink on which can be kept soap, etc., a narrow strip of hooks above the sink for such things as soap dishes, dish mops, dippers, etc.?

Has the work table shallow drawers which hold carving knives, work spoons, knives, forks, etc., and a deeper drawer for extra towels, holders, etc.? Are the skillets, saucepans, kettles, long spoons, etc., on shelves or hook strips near the stove? Have we really furnished the kitchen with all the convenient work tools we can afford? If not, why not?

How few women there are who cannot afford a dover egg-beater instead of a fork; a long-handled soup ladle instead of a cup; sharp knives for carving and slicing; a can opener; a food grinder instead of a hand chopper. Many can afford a bread mixer, fireless cooker and a vacuum cleaner.

The housekeeper can plan her work and save herself a great many steps.

THINGS WE CAN HAVE

Wood or Coal Near Stove. Why not have a box built for the wood or coal, put it on castors, fill it near the door and wheel it right up near the stove? This plan will not only prove to be a great convenience, but will keep a great deal of dirt out of the kitchen.

Supplies and Utensils Handy. The proper arrangement of cooking utensils will lighten the work. Kettles which are used daily should be placed where they can be reached without stooping and without moving other utensils. The potato masher is used at the stove. Hang it on a hook near-by. A wire frame for covers at the side of the range will be convenient. A shelf for salt, pepper and matches, within reach as one works at the stove, will save countless steps. The dishpans should hang near the place for washing dishes. Some housewives insist upon putting them out of sight. In most cases this means a dozen or more extra steps every time they are needed.

This is the daily task of thousands of women who live on the farms—wasted time and energy—which can be saved by installing a home water system

Sharp Knives and a Good Broom. A good paring knife which fits the hand may make the peeling of potatoes a restful task rather than a tiresome one. The farmer has a scoop shovel, a spade, a long-handled and a short-handled shovel, because each one is adapted to a certain kind of work. It seems unfair that his wife should have one poor butcher knife for all processes where knives are needed.

A Sink and Drain. The place of the sink, like that of the stove, is often apparently settled by the builders of the house without reference to the housekeeper's convenience and the position of the other kitchen furniture. It is usually placed with the long side against the wall, but this is not always the best plan.

Some modern houses have the sink near the middle of the kitchen so that it may be used from both sides instead of from one. Insist upon having a drainboard on each side of your sink, one side to be used for dirty dishes, the other for draining. If you are right handed, place the rinsing pan to the left of the dishpan. If left handed place it on the opposite side. This means in washing dishes for the average family a saving of several minutes every time the dishes are washed.

Hot and Cold Running Water. There is really little reason for many farm women going without running water or drainage, as both may be installed at no great expense. If nothing better can be afforded, a large barrel placed outside the kitchen wall and fitted with a pipe and faucet will furnish running water for a large portion of the year. The eaves trough might be connected with the barrel.

Storage tanks which furnish running water throughout the house are being more and more used and are inexpensive and easy to install. On any farm where power is used there is no reason why a good water supply could not be provided for the household.

Light and Fresh Air. The housewife has to spend many hours each day in the kitchen, and sufficient light and ventilation are necessary not only to conserve her health, but to perform her work most efficiently. A window reaching to the ceiling is especially good, as it lets out the hot air as it rises. There can scarcely be too many windows in any kitchen. Small windows above sinks and cupboards add to both lighting and ventilation.

During cold weather good ventilation may be secured by placing a board which is as long as the width of the window and shutting the lower sash upon it. This arrangement will admit air between the two sashes without draft.

High Stool in the Kitchen. Have a high stool in the kitchen and use it. It will seem awkward at first, but cultivate the habit. Sit down to iron clothes, peel vegetables, mix cake, etc.

Screens for Doors, Windows and Porches. Doors, windows and porches should be screened and care should be taken to see that the screens fit tightly, that they are always in place and that doors are not left ajar or held open. If wire screens cannot be afforded, flies can be kept out by cotton mosquito netting tacked over the windows. A piece of netting containing sixteen square yards can be bought for fifty cents.

Things Neat and Clean. Strive to keep the home in good condition; clean, orderly, comfortable, convenient, hospitable; as attractive in furnishing as means will permit; a home which the family will be reluctant to leave, to which neighbors and wayfarers will be eager to come.

A dirty back-yard, unsightly, unhealthy—a place where disease lurks and flies gather

Strive to keep order in the dooryard and ground, so that all who pass in and out will find the day and the way better and fairer.

Flowers and Vines. The morning glory, wild cucumber, Virginia creeper, clematis, wild grape, woodbine or crimson rambler will add beauty and comfort to the porch or around the windows.

Shrubbery set out in the back yard helps wonderfully. Tall shrubs should be set to the back in order not to obscure the view. Use system and not haphazard planting in setting out the plants. Some forms of shrubbery make a better appearance when set out in clumps. Others are best put out singly. Climbing vines, especially roses, are attractive when trained on a trellis.

THINGS TO EAT

"Man Needs a Balanced Ration." Feeding the family is as important as feeding the animals. Food, whether it be for man or animal, has two functions to perform:

First, to build up the body and repair wasted tissue.

Second, to furnish energy for the production of heat and motion.

By observing a few general suggestions it is possible to keep a diet well enough balanced for practical purposes.

Both the growth of the body and its working capacity are dependent upon the amount and kind of food we eat.

Uses of Proteins in the Body. People must have protein substances to build up muscles and tissue. The protein is secured mainly from meats, eggs, milk and cheese, or from legumes and cereals.

The legumes and cereals have such a large per cent of starches and sugars that if we were to depend entirely upon vegetables for our protein we would eat too much of the starches and sugars. Furthermore, the vegetable protein is not so easily digested as that found in animal food. For these reasons it is usually thought best for us to eat a mixed diet and secure a considerable part of our protein from animal food.

Uses of Sugars, Starches and Fats. The best energy-producing foods are sugars and starches in cereals and vegetables. Corn, wheat, potatoes and rice should make up a large part of our diet. The fats are used in the same way as the sugars and starches, but have a much higher energy and fuel value and should be taken in much smaller quantities.

A great deal of fat is secured from butter, oils, olives, fat meats, nuts, etc. Cereals and vegetables contain a small quantity of fat.

Green Foods and Relishes. Fruits, green vegetables and relishes furnish bulk, juiciness, flavor and the needed mineral elements.

Uses of Water in the Body. Water helps to dissolve and distribute the foods in the body and carries off the waste material.

A great deal of water is supplied to us through food material, such as milk and green vegetables, but this is not a sufficient amount. People should drink a great deal of water beside that consumed with meals.

A GARDEN FOR EVERY HOME

Will Lessen Cost of Living—Give Greater Variety— Have Long Rows—Cultivate With a Horse—Plant Early and Late Varieties—Preserve Surplus for Winter Use or Sale—Less Meat and Better Health

Every home should have a garden. Have fruit trees and berry bushes around the sides and ends of the garden. Have the garden long so that you can cultivate it with a horse or mule. Don't just have a little square patch that you will neglect to care for. Cultivate the garden the same as the good farmer cultivates his crops. It will pay you a mighty sight better for the work done in it. The same amount of hoeing necessary to keep the other crops clean will keep the garden clean.

Select early and late cabbage, carrots, turnips, sweet corn, etc. Radishes, lettuce and some of the short-lived vegetables can be planted at different seasons so that there will be a continuous supply. When these vegetables mature, can the surplus for future use. A good garden means one-third of the living.

If you do not know the best varieties to plant, write the Professor of Horticulture at your Agricultural College. He doubtless will furnish you a list well adapted to your locality.

POULTRY PAYS THE GROCERY BILL

Provides Wholesome Food—Destroys Weeds and Insects—Utilizes Waste Products—Small Investment—Little Labor

Next to the garden the most important thing is poultry. Almost every farmer keeps some chickens. The only trouble is, he doesn't make the chickens keep him. Too many of them are roosters. A lot more are old hens that lay but a few eggs and then want to set. Whenever the price of eggs gets high the hens go on a strike. They lay in the henhouse when they please, but too much of the time prefer to lay their eggs out in the weeds. That means the eggs are not gathered until they are stale, and a low price is received for them.

Fifty good business hens are enough to start with. The house for them and their chickens until they are marketed should be about twelve by twenty. The south side should be nearly all covered with burlap or ten-cent cotton with just frame enough to prevent it from whipping too much in the wind.

In front of the house should be a yard, chicken-tight, large enough for the hens to scratch and wallow in. They should be kept shut up in this yard until three or four o'clock in the afternoon. By that time they will have done their day's work and left their eggs where you can find them. Then let them out to hunt bugs and pick grass and find some lime and gravel. For the winter months have some green feed for them.

If possible some of the grain feed during the winter should be furnished in the sheaf. Scratching it out gives them exercise. A load of millet, oats, or cowpeas stacked up well so that a bundle can be thrown in on the floor of the henhouse every day will add immensely to the egg supply. Don't think you have to buy fancy feeds. Grow the feed for the poultry on the farm. Give the hens the scraps from the table. About the only thing necessary to buy for laying hens would be some ground oyster shell. If the boys will shoot a squirrel or rabbit a couple of times a week, skin it and hang it up in the henhouse so that the hens will have to jump a little to reach it, but still can get the meat, there will be no reason in the world for buying meat scraps. Squirrel or rabbit meat makes just as good eggs as beefsteak would if you bought that for the hens. A little poultry handled in this way will pay for all of the supplies an ordinary family will buy. A hundred hens handled in this way will more than do it.

Small Investment. Your investment is small. The commercial poultryman must provide expensive housing and yarding arrangements. You can let your chickens run anywhere— just so you keep them out of the garden during the spring and summer.

Little Labor. Your cost of labor is small; labor is a big item with the commercial poultryman. You can take care of a flock of chickens, and take good care of it without spending a great deal of time. The work fits in with the rest of the farm work.

Cheap Feed. Your feed is cheap. The commercial poultryman buys feed. You raise it, and thus get it at actual cost of production.

Waste Products are Utilized. Your chickens utilize the waste products on the farm. During a great part of the year the hens live almost entirely on what would otherwise be wasted—grass, clover, the gleanings from the grain fields, the surplus garden stuff, and the litter about the barn and feeding pens.

Weeds and Insects are Destroyed. Then, in addition, they eat weed seeds, injurious insects, bugs and worms. It is almost impossible to estimate the help the hens give you in keeping down some of the worst pests.

Record of Eighteen Ohio Farm Flocks. The Ohio Experiment Station made a study of the profits in farm poultry. Records of eighteen typical farm flocks were carefully kept.

These flocks ranged in number from 36 to 370, some were pure bred; others were mongrels. They were kept, fed and tended just as the farmer had been caring for them before the Experiment Station asked him to keep a record.

Here are the results of the investigation. For the sake of comparison we have figured the profit from each flock on the basis of 100 hens in a flock.

The best three flocks yielded, respectively, $247, $154, $153 per hundred hens, while the poorest three flocks yielded $63, $62 and $15, respectively. In no case was there a loss. The average profit per hundred hens of the eighteen flocks was $87.

Poultry Profits. One hundred hens are worth a hundred dollars—just about the price of a good dairy cow. Records of the Cow Testing Associations in Iowa show that the average dairy cow makes a profit of $33.

Poultry not only furnishes the eggs for the family, with some to sell, but it furnishes abundance of the nicest kind of meat, that is always fresh and can be prepared quickly when wanted.

HAVE RURAL DISTRICTS MORE HEALTHFUL

All Natural Advantages for Health and Long Life Are in the Country. Sunshine, Pure Air, Good Food, Comfortable Housing Are Within the Reach of All Country Dwellers, While the Conditions in Life in Other Respects Are Certainly Not Harder Than in the Cities. But these Advantages Must Have Made the Country People Careless of Necessary Precautions.

The rural districts have failed to realize the importance of sanitation, and consequently the rural death rate from typhoid fever, malaria, diarrhoea, etc., is greatly in excess of that of urban districts.

Since outside air in the country is pure, it is assumed that air in country bedrooms and country school houses must be pure.

Since the well has furnished pure water for generations, it is assumed that it will continue to do so.

The unsanitary privy has been in use so long that those used to it overlook its obvious dangers.

The city health officer protects the drinking water of every resident of the city. The city sewer carries the waste of each

family to a common disposal plant or outlet. Hospital care or
rigid quarantine of communicable diseases is enforced by law
in the city, and the greater its natural disadvantages as a dwelling place, the greater is the activity of its health officers.

Health reports from other states indicate that conditions are
about the same in all the states as those in New York.

As these facts become generally known, there is an instant
response from the state granges and other organizations in
rural sections.   These movements will doubtless bring great
results, but it will take time to organize and finance the work.
Meanwhile, the housewife must add to her other duties that of
health officer of the home. She must study personal hygiene,
and the sanitation of the rural premises, including the house

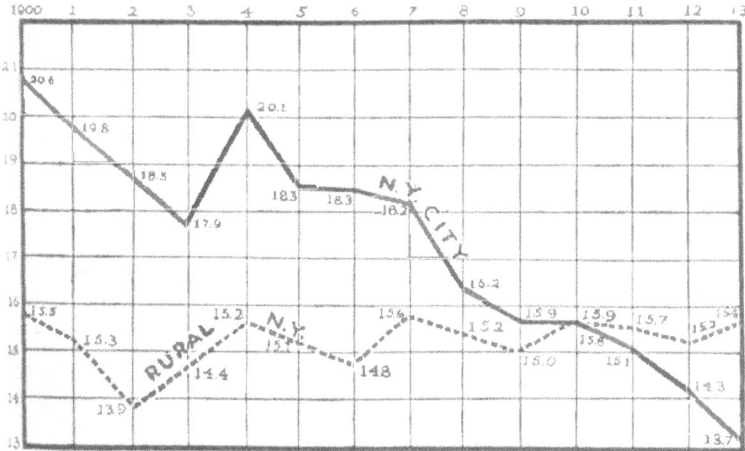

New York Farmers' Bulletin No. 62

**Diagram showing the death rate in New York City compared with the
death rate in rural New York**

and all other outbuildings. She can make living conditions
right for her own family and thus safeguard them and in-
directly her own neighborhood and state. "Sanitation of the
home is the unit which determines the general level of the
sanitation of the state." A high level of both will save lives
valuable to family and state.

Heavy black line on diagram shows decline in death rate in
New York City. Dotted line shows recent increase in death
rate of rural New York (villages below 8,000 population classed
as rural). Note that New York City loses fewer lives in proportion than does the rural portion of the State.

Figures at the top of chart indicate years from 1900 to
1913, inclusive. Figures at side of chart represent deaths from
all causes, per 1,000 of population.

HAVE A SEPTIC TANK

Dispose of All Waste By Septic Tank—Have Indoor Toilets—Use Disinfectants—Outdoor Toilets Permit Breeding of Flies—Pollute Water—Scatter Disease—Health Is Our Greatest Wealth

A simple sanitary system is a most important and necessary feature for every farm home. Without it the health of the family is in danger, however attractive and well arranged the farm house may be in every other respect. It is a matter that should not be postponed to a time when other matters do not press for attention, but should be considered the moment it is needed. The United States Department of Agriculture has issued a bulletin giving a detailed description of a simple sanitary system suitable for the average farm home. It will be sent free to any farmer on application.

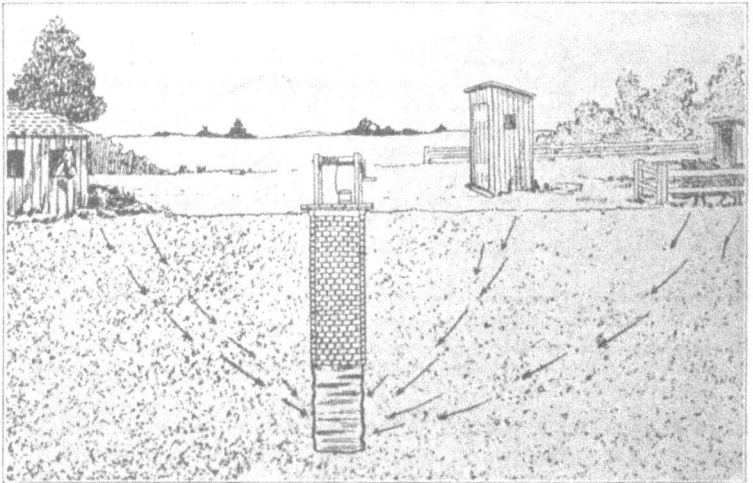

How water becomes polluted in the well. Arrows show course of drainage from the barn yard, manure pile and outhouse

The new bulletin is the result of careful study. Convenience, comfort and economy have all been considered, and they may all be obtained if the suggestions given are practically applied with care and common sense. If the farmer has neither the time nor skill to install the simple system suggested, a reliable plumber, a pump expert or a sanitary engineer should be employed to install a system along similar lines.

The simple sanitary system recommended has four distinct features. It provides for:

1. A pure water supply.
2. Pumping, storage and distribution of water supply.
3. A durable and simple plumbing system
4. A safe disposal for farm sewage.

The above features are described in detail in the department's new bulletin, which consists of forty-six pages and contains thirty-eight figures and diagrams with a number of tables.

YOU DON'T NEED TO HAVE FLIES IN YOUR HOUSE—YOU MAY THINK YOU DO —BUT YOU DON'T

Flies Should Not Be Tolerated—Screen Porches and Windows—Trap the Fly Before He Gets In—Keep the Back Yard Clean—Destroy All Breeding Places—Flies Are More Disgraceful Than Bedbugs

Flies have been tolerated so long without any effort toward their extermination, that many people suppose they must always be tolerated—that it is just one of the things that "must be" and we have to put up with it. They do not realize what a dangerous insect the fly is, and that by screening, trapping it and destroying its breeding places this insect can be exterminated at slight expense. You need not be bothered with these pests—you can:

Screen the porches and windows;
Trap the fly before it gets in;
Keep the back yard clean;
Destroy the breeding places.

A Fly Is More Disgraceful and Dangerous Than a Bedbug.

Dr. Cyrus Thompson, of the North Carolina State Board of Health, says, in a recent bulletin: "As a matter of unprejudiced fact, barring the sting of the bite and the odor of the encounter, the bedbug is much the more eligible companion, whether of bed or of board. But if bedbugs, comparatively cleanly of habit, crawled over your plates, table and food, as the house fly crawls, fresh from the foulest of filth, who could eat or even sit at a table for a moment?

FOOL THINGS WE DO

1—Eat Too Much

2—Keep the Windows Closed

3—Keep Irregular Hours

4—Gossip—Dope—Argue—Complain

5—Leave Poisons Around

6—Go Without Things We Could Have

7—Leave Children Locked in the House

8—Spit on the Floor

9—Have the Blues

10—Be Selfish—Jealous—Suspicious—
      Grouchy

Let Us Give to the World the Best We Have

And the Best Will Come Back to Us

WIFE SAVER AND LIFE SAVER
Let the Engine Do the Work

The engine in the farm home makes possible the combination of a number of labor-saving facilities.

Some of these may be classed as luxuries, but most of them are conveniences and things which the modern farmer can well afford, and they will greatly relieve the burden of household work for the farm wife.

By having a power house for the engine, one can put in a line shaft, connect all the machinery with pulleys, and run them all at the same time with the one engine, or each one can be run separately. (See illustration following title page.)

The engine is becoming practically a jack-of-all trades. It will do away with the weekly turmoil of the wash tub, saving the wife much work and many backaches. It will do the churning, separate the cream, pump the water, grind the feed, saw the wood, and run the vacuum cleaner. It will also furnish power to operate a home lighting plant. It does all these things quickly, easily and with but little expense. The new types of engines suitable for use in the home are made to burn either gasoline or kerosene.

HAVE WHOLE FAMILY PLAN THE WORK

Father, Mother, Children—Give Children an Interest in Something—Chickens, Pigs, Gardens or Crops —Divide Labor and Income—Hold Meetings to Discuss Plans—Success Depends Upon Co-operation

Make the home a co-operative society and let the children have a part in the home making. Teach them early in life that home is not merely a place where they are to have care, comfort and protection, but that for this something must be given, that every member of the family has rights and privileges, and should share in the work and responsibilities.

Give them an interest in something, encourage them to put forth their best effort in everything they do. Give them credit for earning something beside a bare "keep."

There is nothing so valuable as a well developed boy or girl, and nearly all boys and girls are good under the proper treatment. They want work, employment for head, hands and heart—an all-round development; to be taught to think for themselves; to observe, to study out the why and wherefore, and to experiment under intelligent guidance.

Give them a share in all things—your gains and losses. Get their ideas on all subjects, and you will be wonderfully surprised at their abilities. When it comes to making money some of the young folks are in position to teach father and mother in many particulars.

Consult the boys and girls. Take them into your confidence. Have them help solve the problems of the farm and household. Make them realize that they are a part of the machine that runs the affairs of the home. Keep the garden free from weeds. Let the boy help you do it. He will enjoy the result when he sees that his labor has produced something worth while.

Give him a piece of ground to cultivate for himself. This will increase his interest. Let him enjoy the financial returns from his effort.

If you give the boy a pig, remember it is his pig when you sell it. Did you ever know the boy who owned all the pigs and lambs, but whose Dad owned all the hogs and sheep? Professor Holden says, "A boy that measures his wits against a pig's is going to be some man when he grows up."

Interest the girls in the culture of flowers, the raising of poultry, gardening and canning.

Success Depends Upon Co-operation

Endeavor to create a co-operative spirit in the home and in the field management so that the family will live and labor as a unit instead of a chance group of individuals.

The home should be considered a corporation in which every member is a stockholder.

BE SOCIABLE
HAVE BOOKS, MUSIC, GAMES AT HOME

"Get Acquainted with Your Neighbor, You Might Like Him"

There should be choice books, suitable to the tastes and ages of the children, as well as the grown people. Help in the choosing of good books for the home. Much of the future of the child depends upon the influence of the home library.

Take an interest in Community Clubs—Schools—Churches —and Libraries.

If you have no Community Club—help organize one. The men will appreciate the benefit derived from discussing their problems with their neighbors.

The women can study home problems. All phases of home-making—Planning Meals—Dress Designing—Textiles—Home Sanitation—Home Furnishing and Decoration—Household Accounts—Care of the House—Home Nursing, etc.

As a result of the studies in the clubs, women are learning to view housework and homemaking in the right light. Science and art applied to the work do away with drudgery. Through the Community Club arrangements may be made for lecture courses and traveling libraries to be brought to your community. The benefits to be derived are almost unlimited, whether taken from the standpoint of the individual, the community, the school, or the church.

Help make it possible to have Better Roads—Better Schools —Better Farms—Better Health—Better Laws.

"THE HOUSE

IS THE PRODUCT

OF HUMAN HANDICRAFT

A HOME

IS THE CREATION

OF THE HEART"

THE FOLLOWING IS FOR HISTORICAL PURPOSES ONLY
AND DOES NOT REFLECT CURRENT
SCIENTIFIC KNOWLEDGE, POLICIES,
PRACTICES, METHODS OR USES.

BUDGET
YOUR
FOOD SUPPLY

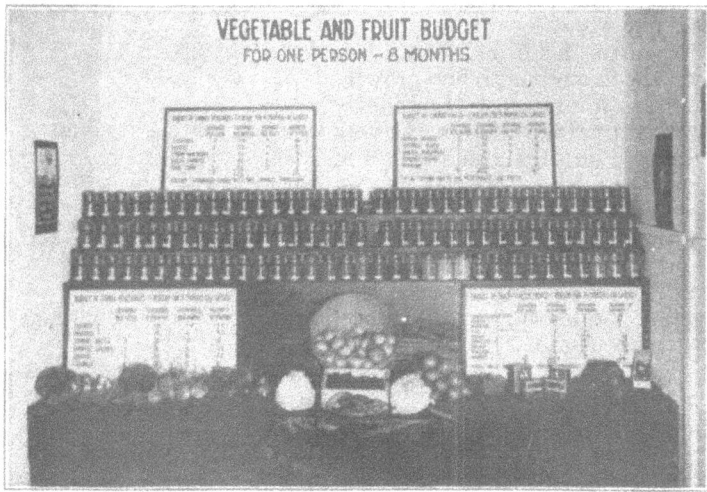

VEGETABLE AND FRUIT BUDGET
FOR ONE PERSON ~ 8 MONTHS

Published 1925 by

INTERNATIONAL HARVESTER COMPANY
INCORPORATED
AGRICULTURAL EXTENSION DEPARTMENT
P. G. HOLDEN, FIELD DIRECTOR
606 SO. MICHIGAN AVE. CHICAGO, ILL.

FOOD AND HEALTH

By Grace Marian Smith

Dr. C. B. Thom, micrologist for the United States Department of Agriculture, is authority for the statement at the head of the chart, "Our Most Important Business Is Feeding the Family."

He also says that selecting and preparing food for the family is so important that it **must not** be left to ignorant, untrained people, no matter how faithful they are nor how good their intentions.

Looks and taste alone are not safe guides in choosing our food. Selecting and preparing the food for the family is work worthy of the best thought and efforts of anyone. Our spirits, judgment, health, and disposition are largely dependent on what we eat.

The chart at the right shows what foods we need to keep ourselves well and to supply energy for doing our work.

Vitamins and minerals are necessary to our health and growth. These are found most abundantly in milk and milk products, the yolk of the egg, and the green leaves of vegetables. Plan the diet around these foods.

A quart of **milk** for children and a pint for grown-ups is the first essential. Milk may be used as a beverage, or in soups, gravies, creamed vegetables, puddings, chocolate or ice cream.

EATING FOR HEALTH

EVERY DAY
NOT LESS THAN
FOR CHILDREN   FOR GROWN-UPS

EVERY DAY
ONE RAW FRUIT OR VEGETABLE
2 SERVINGS OF VEGETABLES
2 SERVINGS OF FRUIT

TWICE A WEEK
A GREEN LEAF VEGETABLE
TOMATOES OR ORANGES

FLOURS & CEREALS FROM WHOLE
GRAINS, EGGS, CHEESE, BEANS, PEAS
NUTS, GELATIN, FISH OR LEAN MEAT

YOU CANT GROW THE FAMILY FOOD
SUPPLY ON A 2×4 GARDEN PATCH

Eat vegetables and fruits twice a day. Learn to like all kinds.

Each day eat one raw food — apples, oranges, cabbage, celery, lettuce, cucumbers, onions, radishes, carrots, turnips, sweet potatoes, melons, or other fruit or vegetable.

Fruits and vegetables also furnish starches and sugars. Sugars from fruits and vegetables are better than too much cane sugar.

Twice a week eat a green-leaf vegetable, such as greens, cabbage, lettuce, green pod beans, or asparagus.

2

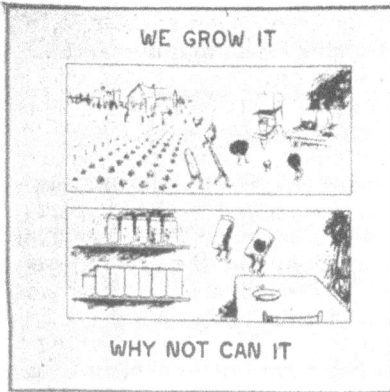

Eat tomatoes or oranges at least twice a week.

Store, can, and dry a supply of vegetables to cover the time when they cannot be obtained fresh from the garden.

Plan to grow and preserve in some way for winter use, enough vegetables to serve some kind other than Irish potatoes twice a day.

A general rule is to plan the food budget for:

Seven servings of canned vegetables each week.

Seven servings of fresh, stored, or dried vegetables each week.

In sections where vegetables for 32 weeks' supply must be canned or stored, making a budget for 36 weeks will allow enough extra for company and some unpreventable waste.

It is an advantage to have some of these foods canned for use even in the summer months. It is quicker, easier, and cheaper to cook—say, 30 meals of string beans or fried chicken at one time, than it is to cook string beans or fry chicken 30 separate times.

Canned foods are ready to serve when wanted: for Sunday dinner, when unexpected company comes, when we go home tired and late after a fair or short course; easier when threshing-day comes.

Creamed vegetables, chicken a la King, meat and vegetable stews may also be canned for quick service when needed.

Use whole grains for flours and cereals. The best flavor, minerals, proteins, and oils of any grain or fruit lie just under the skin and around the stone or core which contains the germ.

The minerals, oils, flavors, and proteins have been taken away from "bolted" wheat flour, hominy from which the hull and germ have been removed, polished rice, and similar foods, leaving nothing but starch. We get starch in many foods. We need the other food elements which we take out of the grains and feed to the stock.

Eggs, cheese, beans, peas, nuts, gelatine, fish, and lean meat furnish

3

building foods. Some of these, with butter and fat meats, also furnish additional fat.

Everyone should be taught how to select food for health.

The chart, "Budget Your Canned Food Needs," shows a suggested budget of the canned foods needed to supply a family of five for the eight months of the year when gardens in some sections are not producing.

Canned food has the advantage of being ready to use when it is wanted and of retaining some vitamins which may be lost in the more mature product which is stored.

Raw Vegetables Needed to Make 1 Quart Canned

Corn—12 to 20 ears.
Beans, string—1½ lbs.
Beans, lima—1½ to 2 qts.
Cabbage—2 medium heads.
Cauliflower—2 medium heads.
Carrots—2 to 2½ lbs.
Beets—2 to 3 lbs.
Turnips—2 to 3 lbs.
Peas, in pod—3 to 4 qts.
Spinach—1 pk.
Squash—2 large.
Pumpkin—1 medium cans about 4 qts.
Tomatoes—1 bushel cans 12 to 15 qts.

Servings Per Pound of Stored, Brined, and Dried Vegetables

Root vegetables weigh about 50 pounds per bushel, and average 4 servings per pound; Squash, 1½ servings per pound; Cabbage, raw, 7 to 8 servings; cooked, 4 to 5 servings; Brined vegetables, 6 to 8 servings; Dried vegetables, 6 to 8 servings; Kraut, 4 servings per quart.

4

Food Preservation Budget Summary

This estimate is for one person and is based on the use of two servings of fruit, one serving of potato, and two servings of vegetables other than potato daily throughout the year.

VEGETABLES

	Total Fresh	Total Stored	Total Canned
Greens			
Spinach			Av. of 7 lbs.
Lettuce	60 servings	10 lbs. greens	or 7 pts. of all
Beet Tops	or 15 lbs.	may be salted	greens
Chard			
Cabbage	5 lbs.	15 lbs.	
Carrots, Squash and Pumpkin	20 lbs.	30 lbs.	7 pts.
String Beans and Asparagus	$\frac{1}{4}$ bu. $1\frac{1}{2}$ bu.		10 pts.
Green Peas, Green Lima Beans Corn	$\frac{1}{3}$ bu. $\frac{2}{3}$ bu. or 30 ears	3 pts. dried	6 pts.
Turnips and Rutabagas	10 lbs.	20 lbs.	
Onions	$12\frac{1}{2}$ lbs.	25 lbs.	
Beets and Parsnips	15 lbs.	20 lbs.	7 pts.
Potatoes, Irish or Sweet	45 lbs.	135 lbs.	

FRUITS

	Total Fresh	Total Stored	Total Canned
Tomatoes	22 lbs.	42 tomatoes salted	8 qts.
Prunes, Raisins, Figs, and Dates		11 lbs.	
Apples	48 apples	143 apples	
Cherries	6 qts.		
Rhubarb	20 servings		
Berries	9 qts.		4 qts. altogether of
Plums	$\frac{1}{8}$ bu.		any one or
Peaches	$\frac{1}{8}$ bu.		more
Pears	$\frac{1}{8}$ bu.	12 lbs. pears	
Melons	19 servings		
Grapes	12 "		

5

Garden Space Required Based on Food Preservation Budget

By R. M. ADAMS, Assistant Extension Professor, Vegetable Gardening
New York State College of Agriculture, Cornell University, Ithaca, N. Y.

Vegetables Supply for One	Amount of Product	Feet of Row
Greens	34 lbs.	Spinach, 250 feet or Chard or New Zealand Spinach, 50 feet, or proportion of each
Cabbage or Collards	30 lbs.	17
{ Peas or Pole Limas	1 bu.	75
{ or Corn	55 ears	55
Carrots	50 lbs.	65
Squash or Pumpkin	14 lbs.	1 or 2 hills
String Beans	½ bu.	27
Asparagus	5 lbs.	10
Turnips and Rutabagas	30 lbs.	40
Onions	35 lbs.	65
Beets and Parsnips	50 lbs.	90
Potatoes, Irish or Sweet	180 lbs.	400
Tomatoes	55 lbs.	64

Multiply these figures by the number of people in your family.

You may wish to substitute okra, salsify, kale, cauliflower, celery, eggplant, radishes, cucumbers, or melons for part of some of the items given. In season we shall want fresh lettuce, cucumbers, and other relishes.

Don't plant all the lettuce, corn, etc., at one time. Plan a succession of plantings. Plant early and late varieties.

Probably we shall always store some cabbages, beans, potatoes, root crops, and some other foods. We may dry some corn and smoke and brine some meat, because we like their flavors. Canned foods taste like fresh and are ready to serve.

On the front cover is a photograph of a display made by the Nebraska Agricultural College. This includes canned, dried, and stored foods, sufficient for one person for eight months.

6

Home Canning Guide
One-Period Cold-Pack Method

Water, heat, and all equipment including cans and jars should be ready before starting. Jars and covers should be clean and hot.

Select fresh, firm, sound products.

Clean, grade, pare, slice, or otherwise prepare.

Scald or blanch vegetables and hard fruits. It is not necessary to heat soft fruits and berries before packing.

Dip product quickly into cold water. Pack at once in clean hot containers.

Pour hot syrup or boiling water over fruit.

Add one teaspoon salt per quart to vegetables and fill can or jar with boiling water.

Wipe neck of jar with clean cloth before putting rubber in place.

Use good rubber rings.

Place scalded rubber and cap in position at once.

Partially tighten tops on jars. Seal tin cans completely.

Sterilize as given in the time-table. Do not begin to count time until the water is at a jumping boil.

Remove can from canner, seal completely, invert to test joints for pinhole leaks.

Plunge tin cans into cold water to cool.

Cool glass jars as quickly as possible and avoid drafts.

Label, wrap and store.

SUGGESTIONS

Cherries, blackberries, raspberries, all red fruits, gooseberries, pumpkin, beets, squash and sweet potatoes lose color if canned in unlacquered tin.

Use lacquered cans or glass jars for very acid products.

Rhubarb should always be canned in glass. It contains a very strong acid which will affect even the lacquered tin.

Corn, beans, peas, pumpkin, squash, and sweet potatoes swell in cooking. Do not fill the cans too full. These products should not be packed so tight that the heat can not reach the center readily. All these products require a high degree of heat and a long period of sterilizing.

Can corn as soon as possible after picking it.

Storing: Do not put tin cans in the hot sun or a hot room nor pack them together too close when they are taken from the cooker, or they will hold the heat and overcook.

Do not store in a damp place.

Meats may be cooked in the can or seared and partially cooked before packing. Can dry or add water as desired.

Planning the year's food supply for the family is at least as important as planning the feed for the stock.

The whole family must help.

7

TIME-TABLE FOR CANNING

Use the time given under the type of outfit you are using.

PRODUCTS	Preparation	Hot Water Bath Outfits at 212°	Steam Pressure Cooker
Fruits			
Apricots.............	Scald	25 Min.	
All berries...........	Cold Pack	15 "	
Cherries.............	Cold Pack	20 "	
Currants.............	Cold Pack	15 "	
Grapes..............	Cold Pack	15 "	In using pressure
Peaches.............	Scald	25 "	kettle for canning
Plums..............	Cold Pack	20 "	fruits and tomatoes,
Rhubarb Sauce......	Hot Pack	5 "	keep the pressure
Apples.............	Blanch 1½ Min.	12 "	at 1 or 2 lbs. and
Pears..............	Blanch 1½ "	25 "	process for the same
Pineapple..........	Steam 10 "	25 "	length of time as
Quince.............	Blanch 1½ "	25 "	given for the hot
Figs...............	Blanch 15 "	40 "	water bath outfits.
Some Specials			
Tomatoes...........	Scald	40 Min.	
Eggplant...........	Preheat	1½ Hrs.	40 Min at 10 lbs.
Pumpkin...........	Precook	3 "	2¼ Hrs. at 10 lbs.
Squash............	Precook	3 "	2¼ " at 10 lbs.
Corn..............	Preheat	3½ "	1½ " at 10 lbs.
Hominy............	Preheat	3 "	1 Hr. at 10 lbs.
Greens and Vegetables			
All Greens..........	Shrink	3½ Hrs.	1½ Hrs. at 10 lbs.
Asparagus..........	Blanch 4 "	2 "	30 Min. at 10 lbs.
Beans, string.......	Preheat	2½ "	40 " at 10 lbs.
Beans, lima.........	Preheat	3 "	40 " at 10 lbs.
Okra..............	Preheat	2½ "	40 " at 10 lbs.
Peas..............	Preheat	3½ "	40 " at 10 lbs.
Brussels Sprouts.....	Preheat	2 "	40 " at 10 lbs.
Sauerkraut.........	Cold Pack	30 Min.	15 " at 5 lbs.
Cauliflower.........	Blanch 3 Min.	1 Hr.	40 " at 5 lbs.
Beets.............	Scald	2 Hrs.	40 " at 10 lbs.
Carrots............	Scald	2 "	40 " at 10 lbs.
Sweet Potatoes......	Scald	3 "	70 " at 10 lbs.
Meats			
Beef and Pork.......	Cold Pack	3 Hrs.	70 Min. at 15 lbs.
Poultry............	Cold Pack	3½ "	80 " at 15 lbs.
Fish..............	Cold Pack	3 "	80 " at 15 lbs.

*Soups—In canning vegetable mixtures for soup, use the time required for the vegetable which takes the longest cooking. For soups made with meat or meat broth, cook the meat first, add the chopped vegetables and boil 10 minutes. Pack in cans or hot jars and cook 2 hours in Hot Water Bath or Steam Cooker or 50 minutes in Pressure Cooker at 10 pounds pressure.

*Asparagus, Corn, Greens, Lima Beans and Peas should be canned only in pint glass jars or in No. 2 tin cans

*Temperature in Steam Cookers varies, never reaching more than 211½°. Use time-table for Hot Water Bath outfit, or increase the time slightly.

*This time schedule is based upon the 1-quart pack and upon fresh products at altitudes up to 1,000 feet. For higher altitudes, increase the time 10 per cent for each additional 500 feet.

Three Studies in Soil

1. Testing Soil for Acidity

Many forms of plant life will not grow in a sour soil. There are millions of acres of acid soil in the United States.

Sometimes a field which does not make a good home for plants, will grow large crops if lime is added to the soil to sweeten it.

Here are two ways to find out whether soil needs lime or not:

Test No. 1. Cover a piece of blue litmus paper, which you can get at any drug store, with some of the moist soil. Let it stand a few minutes. If the paper turns pink, it means there is acid in the soil.

Test No. 2. Pour a small quantity of muriatic (hydrochloric or "tinner's") acid on the soil. If there is lime in the soil, the acid will form tiny bubbles.

Such soil does not need to have lime added to sweeten it, but the addition of lime may make larger crop yields.

What Lime Will Do

The addition of lime: Improves the texture of the soil. Breaks down the soil particles. Lets in air and moisture. Liberates nitrogen and other plant food.

Makes sour soil sweet—enlivens it—that is, makes conditions favorable to plant growth by giving life to the soil bacteria which make it productive.

2. Saving Soil Moisture

One reason we cultivate ground is to make a fine, soft bed for the seed.

Another reason is to destroy the young weeds as they appear.

The most important reason is to help keep the moisture in the ground.

When rain falls on the earth, some of it runs off in little rivulets; but some of it soaks down into the ground until it strikes a layer of rock or hard clay which will not let the water through.

When it can soak no further the water may form a little river and run underground, sometimes gushing out somewhere in the form of a spring; or it may soak upward through the soil until it reaches the surface of the ground.

The soil a storehouse. The soil is a storehouse for plant food. It is also a storehouse for moisture.

Sometimes people complain that they do not have rain enough to make crops grow, but if they will not let the moisture evaporate from the soil they will have all they need to raise a crop.

How dust helps. You may wonder how cultivation keeps moisture in the soil.

Throw a dipper full of water into a road or street where the dust is very deep. Notice how the water forms little balls or bubbles and does not soak into the dust. This is because the fine dust will not let the water through.

If water will not soak down through dust it will not soak up through it.

Fig. 1. Placing the powdered sugar on the cube, or in other words, adding the soil "mulch"

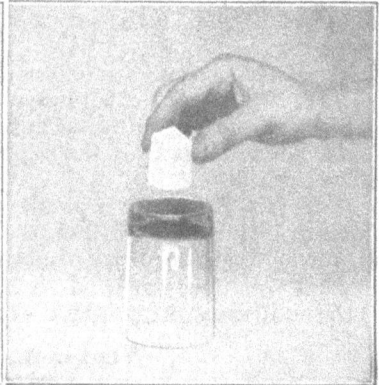

Fig. 2. Placing the cube of sugar on the concave bottom of the glass containing the solution

The sugar experiment. Take a cube of sugar—"loaf" sugar is best, but a lump of ordinary sugar will do — and cover the top of it with a quarter of an inch of fine, powdered sugar. (See Fig. 1.)

Turn a drinking glass, with a curved bottom, upside down. (Or the bowl of a spoon may be used.) Partly fill the concave bottom of the glass with water that has been colored with ink. Place the cube of sugar, bottom down, in the colored water. (See Fig. 2.)

Fig. 3. The cube ten seconds after it was placed in the solution

Notice how quickly the water soaks up through the loaf sugar until it reaches the top, but when it gets to the powdered sugar it stops. (See Fig. 3.)

It will soak through the inch of loaf sugar in about 10 seconds, but it will take an hour or more for it to soak up through the quarter inch of powdered sugar. In this experiment the loaf sugar represents the soil, the powdered sugar represents the dust.

The dust mulch. When we put a layer of fine soil dust, or "dust mulch" as it is called, all over the top of our field, we keep the water in the ground for the plant roots to drink up.

We can make such a mulch with a harrow or a drag, or, in a small plot, with a hand rake, and if it bakes down we have only to disk it over again.

Soil should be stirred every week or 10 days, and always after a rain.

Sometimes soil bakes hard and cracks open. This is because all the water has evaporated.

Soils should never be allowed to crack in this way. The ground should be kept stirred on top so as to keep the moisture in the ground.

This story and the previous one are told in the I H C Slide Lecture Book, "A Fertile Soil Means a Prosperous People," published by the Agricultural Extension Department of the International Harvester Company. Price 6 cents.

3. The Study of Nodules

Nodules are tiny, knotty-looking lumps attached to the root hairs of some plants.

Nodules are not found on the roots of corn, or wheat, or oats, nor on timothy, or cabbages.

They are likely to be found attached to the root hairs of any healthy, vigorous-looking legume. A legume is a pod-bearing plant.

Alfalfa, all of the clovers, vetches, beans, and peas, and hibiscus, locust, and catalpa trees, are common examples of leguminous plants.

Sweet clover is especially likely to have a great many nodules on its roots.

Sometimes we find no nodules on the roots of one plant, but find them on the roots of another plant growing right beside the first one.

In such cases the plant with nodules on is likely to have a better color and be

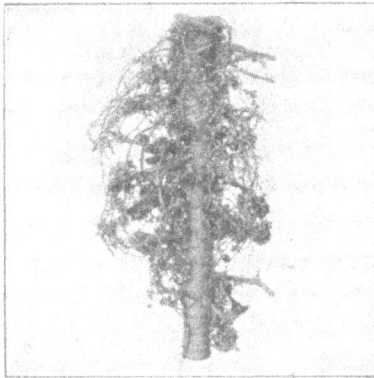

An unusually well-developed growth of nodules on a sweet clover root

a much more vigorous looking plant than the other one.

Nodules are not found in hard, baked soils. Clay, for example, often packs solidly together until it is necessary to add some humus or lime to it, when it may become a good home for bacteria.

Nodules are most numerous in the spring, and on young plants just starting to grow, and where the ground is rich, sweet, and moist, but not wet.

You may have seen nodules on the roots of clover, and thought they were worm's eggs, or a growth caused by disease.

They are the homes of wonderful little organisms.

Nodules are the home of bacteria. Different kinds of bacteria live on different legumes but all of them do the same work.

Alfalfa, sweet clover, and bur clover have the same kind of bacteria on their roots. Crimson, red, white alsike, and California clovers all have the same kind of bacteria, but it is a different kind from that which is on alfalfa.

Cow peas have another kind of bacteria, soy beans another kind, peanuts still another kind. Garden peas, Canadian field peas, and most of the vetches all have the same kind of bacteria —a still different kind from any of those mentioned before.

Some trees bear their seeds in long pods. See if you can find any nodules on their rootlets. Remember, you are not likely to find nodules where it is very dry.

Some forms of bacteria collect nitrogen from air.

To make a good home for plants, soil should be loose and porous, with plenty of air space.

It is from this air in the soil that the bacteria gather the nitrogen.

Nitrogen is one of the most important and most costly elements in plant food.

When we buy fertilizer, we pay a high price for the nitrogen in it.

Nitrogen forms four-fifths (by volume) of the air around us, but we have not yet learned to collect this nitrogen so that it can be used commercially.

How to gather nodules. Dig up the legumes, taking care not to crush or injure the roots, shake them lightly to shake off the dirt, and look for the white, knotty bunches, or single, small nodules.

If you do not find them on the roots of the first plant you dig up, try another growing in soil which is richer and more moist.

You can keep them as long as you care to, by putting them into a wide-mouthed bottle, or a glass fruit jar, and covering them with water, to which you have added four tablespoons of formalin to each quart of water.

Put them in just as they are, attached to the root.

Remember that the white, bulbous formations are the **nodules** in which the bacteria live, **not the bacteria themselves.**

Alfalfa will not grow well where the soil lacks its kind of bacteria. Sometimes people take soil which is full of these bacteria and sprinkle it over a field which they are going to seed to alfalfa, in order to get the bacteria started there.

In the I H C Chart Lecture Book, "Alfalfa on Every Farm," (price 6 cents) there is a story of how to inoculate soil with bacteria.

Published By

INTERNATIONAL HARVESTER COMPANY
INCORPORATED
AGRICULTURAL EXTENSION DEPARTMENT
606 SO. MICHIGAN AVE. CHICAGO. ILL.

AE 50 M. 5M—9-23-30. Printed in U. S. A.

THE FOLLOWING IS FOR HISTORICAL PURPOSES ONLY
AND DOES NOT REFLECT CURRENT
SCIENTIFIC KNOWLEDGE, POLICIES,
PRACTICES, METHODS OR USES.

GROW A VEGETABLE GARDEN

WE MUST FEED OURSELVES

GROW A VEGETABLE GARDEN

**THERE SHOULD BE A GARDEN
FOR EVERY
HOME IN AMERICA**

**WE MUST PUT OUR IDLE
LAND TO WORK**

WE MUST FEED OURSELVES

By J. H. PROST
Agricultural Extension Dept.
International Harvester Company of N. J. (Inc.)

Published and Copyrighted 1918 by
INTERNATIONAL HARVESTER COMPANY
OF NEW JERSEY (INCORPORATED)
AGRICULTURAL EXTENSION DEPARTMENT
P. G. HOLDEN, Director
HARVESTER BUILDING, CHICAGO

AE 451-8 15-18

WHEN TO PLANT

1. Have a Cold Frame. It will lengthen the garden season. Grow lettuce, radishes and other vegetables for early use. Plant cabbage and tomato seed in your cold frame bed and transplant the young plants to the garden.

2. Plant potatoes, onion sets, radishes, black seeded sweet peas as soon as the ground can be spaded or plowed in the early spring. These are very hardy plants.

3. Plant wrinkled peas, beets, Swiss chard, lettuce, cabbage, cauliflower, etc., when the peach and pear are in bloom.

4. Plant corn, string beans, melons, cucumbers, squash, pumpkin when the apple trees begin to blossom.

5. Peppers, okra, tomatoes, lima beans are tender plants and should not be planted until all danger of frost is over—when the hard maple is in full leaf.

Farm gardens should be planted in long rows and cultivated with a regular horse cultivator. Don't plant a little patch of ground near the house, where the chickens will harvest the crop. Many farmers lay out the garden near the house where it will be handy for **mother to care for.** Don't add to the work of the farm home with a make-shift "Slacker" garden of this kind. Have a real garden out where it can be properly cultivated with a team. In the large cities and small towns, put the back lots to work. Do the best you can; but no matter where you plant the garden, care for it, cultivate it, keep down the weeds—get the best out of your work; and whatever you do, don't make the mistake of wasting vegetables, after you have grown them. Can, dry, or store the surplus.

FOOD OUR GREATEST NEED

O UR Government is appealing to us to help win the war—to produce, to conserve—to meet squarely and unfalteringly the problems as they are—and it is our sacred and patriotic duty to answer that appeal. We are impressed with the great importance of food. Without food we cannot fight. Food is the world's great problem. 'We must put our idle lands to work. Every idle acre in America which can produce food and which is not made to produce it, is a direct blow at American liberty. Weeds and waste land are enemies in disguise to discount the efforts of our boys who are fighting abroad and suffering to maintain the cause of human rights. We cannot all fight in the trenches, but we can help in other ways.

The farmers have been called upon to grow more grain and live stock. They have responded. There is a speeding-up in every line of human endeavor. There are wonderful opportunities for us all to help. There should be a garden for every family. Children can help in town and country. Growing vegetables is the most healthful, most useful, most profitable play in which children can engage.

Raise both early and late varieties of vegetables, and plant at frequent intervals, so that you will have a continuous supply throughout the season.

When we grow fruits and vegetables we will save food. A garden will provide about one-third of the supplies for the table.

> ## WHY NOT GROW A GARDEN
>
> LESSENS COST OF LIVING
> PROVIDES GREATER VARIETY
> VEGETABLES ARE CHEAPER
>   CLEANER FRESHER
> WE LIKE THEM BETTER
> WE GROW THEM OURSELVES
> PUT UP SURPLUS FOR WINTER
>
> IF WE DON'T GROW THEM
>   WE GO WITHOUT THEM
> MORE GARDENS
>   MEAN BETTER HEALTH

3

HOME GROWN FOODS BEST   Home grown vegetables are fresher and cheaper than those secured from the nearby grocery store or huckster. You will have a greater variety of food and eat less meat. About 80 per cent of the medicines sold in drug stores are laxatives, and continued use of these drugs is injurious to health. More fruit and vegetables and less meat make laxatives unnecessary.

PLAN FOR GARDEN 18x20 FEET

Seeds for This Garden Will Cost About 75c Retail

18 X 20 FT.

```
—BEETS————————————————————————
—CARROTS————————————————————————
—LETTUCE————————————————————————
—ONIONS————————————————————————
—PARSLEY————————————————————————
—BEANS————————————————————————
—RADISHES————————————————————————
—TURNIPS————————————————————————
—SWISS CHARD————————————————————
—ENDIVE————————————————————————
—LEEK————————————————————————
—KOHL RABI————————————————————
—2ND SOWING RADISHES————————————————
—CABBAGE————————————————————————

—CUCUMBERS————————————————————————
         FLOWERS
```

ROWS 1 FT. APART

2 FT.

2 FT.

A Collection for Garden 18x20 Ft.

1 PKT. EACH		1 PKT. EACH	
Beet	1/4 oz.	Radish	1/4 oz.
Carrot	1/4 oz.	Turnip	1/4 oz.
Cucumber	1/4 oz.	Swiss Chard	1/4 oz.
Lettuce	1/4 oz.	Cabbage	1/16 oz.
Onion	1/8 oz.	Endive	1/16 oz.
Parsley	1/4 oz.	Leek	1/8 oz.
Beans	3 oz.	Kohl Rabi	1/16 oz.

HOW TO GET A START

Manure the Land—Plow or Spade Deep—Make Fine, Firm Seed Bed—Plant Early and Late Varieties—Cultivate Often to Save Moisture and to Kill Weeds.

FALL WORK— PLOWING Start your garden in the fall after crop has been harvested. Manure the land, using about two good wagon-loads for 400 square feet of ground. Wood ashes or bone meal are good. Plow or spade your garden in the fall; it will expose the insect pests. Freezing will break up the soil, make it loose and porous, so that it will retain more moisture and allow the air to act on the soil.

> ## HOW TO GROW A GARDEN
>
> MANURE THE LAND
> PLOW OR SPADE DEEP
> MAKE FINE FIRM
> SEED BED
> PLANT EARLY AND
> LATE VARIETIES
> CULTIVATE OFTEN
> TO SAVE MOISTURE
> TO KILL WEEDS
> MANURE AND PLOW LAND
> NEXT FALL FOR THE
> FOLLOWING SEASON

CATCH CROP A further advantage may be gained by planting rye or clover in the fall after plowing, to be plowed or disked under in the spring. It takes several seasons to work up an ideal soil.

How to Make a Plan

WINTER WORK Make a diagram and planting plan to guide and direct your work throughout the season. Draw the plans on paper and study them.

PLANS ELIMINATE MISTAKES With a plan you will make fewer mistakes and get better results. In making your plan consider the following points:

1. Plan your planting rows lengthwise and long. Cultivate with a horse if possible. Have no cross paths and only one path running lengthwise along the border or edge of garden.

2. If you must do hand cultivating, plan your garden in squares, with paths running crosswise if desired. Plant your vegetables closer in the rows and rows nearer together.

3. Locate your permanent crops, such as asparagus, rhubarb, berries, etc., where they will not be interfered with. (See Plan, Page 7.)

4. Locate a hotbed on north or west line of garden—give it a south or south-eastern exposure—this gives the best sunlight.

5. Locate the compost, garbage, or rubbish pile back of your hotbed, out of sight.

6. Plan for planting your celery, onions, and late cucumbers in low places, and early vegetables in the high, warm, and dry soil.

A GARDEN FOR
          EVERY FAMILY
——
PROMOTES THE HABIT
          OF THRIFT
HELPS SAVE MONEY TO
          BUY A HOME
INTERESTS THE WHOLE FAMILY
HELPS WHEN OUT OF WORK
MAKES A BETTER
          NEIGHBORHOOD
KEEPS BOYS AND GIRLS
          OFF THE STREET
——
WE CAN'T GROW BOYS
          AND GIRLS
WHERE WE CAN'T
          GROW PLANTS

7. Plan for a succession of crops— do not follow with a second planting of the same vegetable. Follow early peas with celery and late peas. Follow early cabbage or potatoes with late beans or corn.

8. Consider appearance — plant the highest growing vegetables like corn at the rear of the garden, and the lower growing in front or nearest the house. Such an arrangement will be most attractive and form a screen at the rear of the garden. Do not let the high-growing plants shade the lower-growing ones.

9. The plan should show the distance between rows, the number of plants in the rows, space between plants, and names of varieties.

School children, especially those in the upper grades and high school will take a keen interest in making these garden designs, planting plans, and specifications. Garden plans should be made during the winter months.

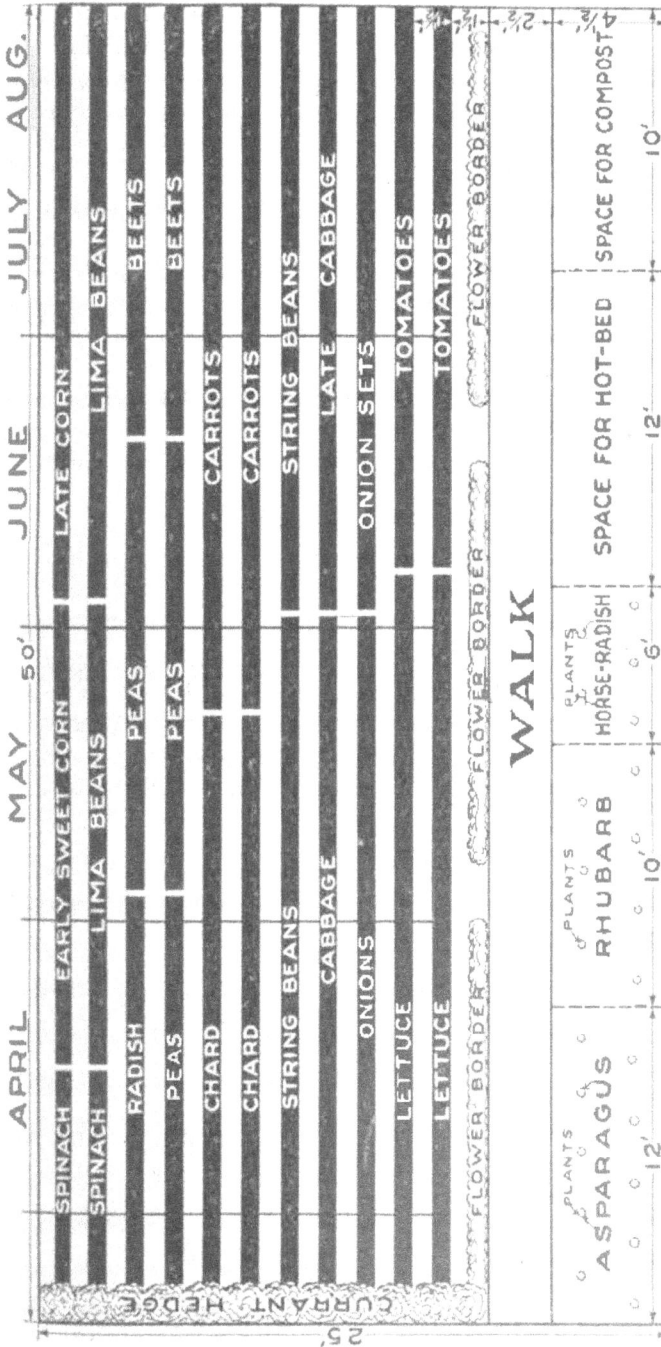

A Vegetable Garden Plan: Showing currants to divide garden from yard, or by continuing hedge around garden you can enclose entire garden. Asparagus, rhubarb and horse-radish are permanent growers, shown planted on one side of walk where these beds will not be disturbed.

The hotbed and compost pile are also permanent features of a successful garden, this plan shows a strip 4½ feet wide, running full length of garden and 2½ feet for walk to be permanently established.

The divisions by months indicate what varieties should be planted during the month, and the flower border is for decorative purposes.

The young folks can make them in school, or with the parents at home. Changes or improvements can be made each year.

In making the plans for school, factory, home, and vacant lot gardens, follow these steps and apply them to local conditions:

1. Make a survey, or measure the length and width of the area to be used for a garden.

2. Plot area on paper, one inch on paper being equal to one or two feet of garden area.

3. Locate your fence, compost heap, hotbed, paths or garden walk and all other permanent features of the garden.

4. Locate your permanent vegetable beds; as asparagus, rhubarb, and horseradish.

5. Mark off your vegetable rows; let each dot indicate a plant.

6. Name each variety, date to plant, and number of plants.

Things You Will Need

THINGS NEEDED Make up an itemized account of the things on hand and things needed.

1. Plans: Garden plan, hotbed plan, planting plan.

2. Tools which are useful: Wheelbarrow, rake, spade, shovel, hatchet, mallet, hoe, tape line, long straight stick marked off in feet and inches, stakes, markers, labels, garden line, dibble, trowel, weeding hook and hand weeder. These tools are not all needed at once, some may not be needed at all, but all are useful in garden work.

CARE FOR YOUR TOOLS Have a place for every tool and keep every tool in its place. It is annoying to find tools missing and out of repair when you most need them, or that they are lost and new ones must be purchased.

Repair your tools during the winter, and have them in readiness when needed.

3. Amount and kind of fertilizer to

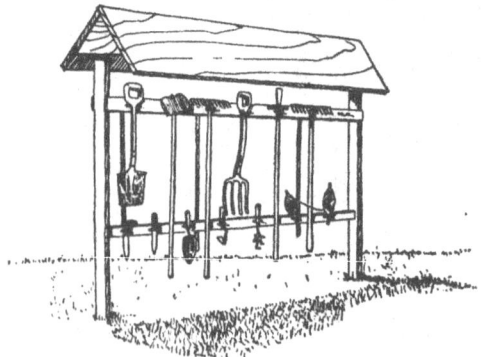

A Tool Rack

be used: Compost, manure, poultry droppings, wood ashes, pulverized sheep manure, and commercial fertilizers.

4. Seeds: The number needed of each variety and cost of same.

How to Prepare the Soil

SPRING WORK Drainage is always very important. Leave no depressions and pockets where water can stand.

Form planting beds with convex surface; this gives surface drainage. Heavy clay soils need tile draining. Sandy soils may be sufficiently drained with side or border ditches. A deep trench or ditch filled with rock, broken crockery, or glass, running along lowest level of garden will act as a suitable drain for small gardens. The average city lot can be well drained with the ditch drainage system.

FERTILIZING A good soil must contain nitrogen, potash, phosphorous and lime. For vegetable growing these plant foods should be applied to the soil each season.

Leaf plants, as cabbage, lettuce, chard, etc., need nitrogen; the vine plants, as tomatoes, beans, cucumbers, must have more phosphorous and potash; and the root crops, as potatoes, turnips, carrots, etc., need potash.

STABLE MANURE IS BEST Well rotted stable manure is about the best and cheapest garden fertilizer. Secure all you can. Do not let the city garbage wagon carry away the stable manure provided in your neighborhood. Put it on your garden.

Good Fertilizers Can be Home Made

HOW TO MAKE A COMPOST PILE Make a compost pile near or in your garden. Compost is made by piling a layer of soil on top of a layer of manure; then a layer of leaves or any organic matter, another layer of soil and some more organic matter, such as straw, stable manure, leaves, grass clippings, plant and meat wastes, ground bones, tree and shrubbery clippings, all piled and mixed and allowed to decompose during Winter. This pile should be turned over twice during the winter months and applied to the garden in the early spring. Compost is one of the best lawn fertilizers and helps vegetable growth when scattered over the ground around young plants.

OTHER   Poultry droppings are very good but should always
MANURES   be applied in the fall.   These are strong in nitrogen.
All manures should be well rotted before applying
to the soil.   Raw or fresh manures should never be applied so as
to come in contact with young seedling plants.

**Make compost by forking over a pile of materials
to mix thoroughly.**

COMMERCIAL Commercial fertilizers are sold under guar-
FERTILIZERS anteed analysis.   These are nitrate of soda,
acid phosphate, dissolved bone, pulverized
sheep manure, etc.   No definite rule can be given for kind, quantity and quality to be applied as this varies with the crop to be
grown and land to be fertilized.

**HUMUS ESSENTIAL
TO GOOD SOIL**

1 - GIVES LIFE TO SOIL
2 - MAKES HOME FOR BACTERIA
3 - KEEPS SOIL LOOSE AND WARM
4 - LETS AIR AND WATER
                INTO THE SOIL
5 - PREVENTS WASHING, BAKING
                AND PACKING
6 - BUILDS SOIL — ADDS NITROGEN
7 - HELPS MAKE OTHER
                PLANT FOOD AVAILABLE

Use about four pounds of
the well balanced vegetable
fertilizers to 100 square feet
of garden area.

WOOD   Don't waste wood
ASHES   ashes.   They contain
a large per cent of
potash.   Spread them over
your garden soil.

LOAMY   Vegetables will grow best in loamy soils.   There
SOIL   are sandy loams and clay loams.   Decomposed vegetable and animal matter when applied to the soil
will produce a loamy texture.   Heavy clay soil needs repeated
applications of manure.

SOUR   Sour soils may be sweetened by broadcasting applica-
SOILS   tions of pulverized limestone, about 15 pounds to 25
            square feet of soil.   Wherever a mossy growth appears
on the ground surface of your garden, make a test for sour or
acid soil.

LITMUS   This test is made by taking half a glass of soil,
TEST FOR   add water to make a muddy solution.   Into
SOUR SOIL   this mud place a piece of blue litmus paper, which
                  you can get at the drug store.   If the litmus
paper turns red the soil is sour and should be treated with lime.

FOR   Soils that bake and crack on the surface need liberal
BAKED   applications of lime; about 25 pounds to 25 square
SOILS   feet, every three years.   Get this lime at the seed
             store, farm implement store, or building materials
company.

BEST   Loose, rich soils that hold the moisture and plant food
SOILS   in solution, allow the free passage of air, and are free
            of acid are the best.
    These are produced by proper drainage, fertilizing, liming,
and cultivation.

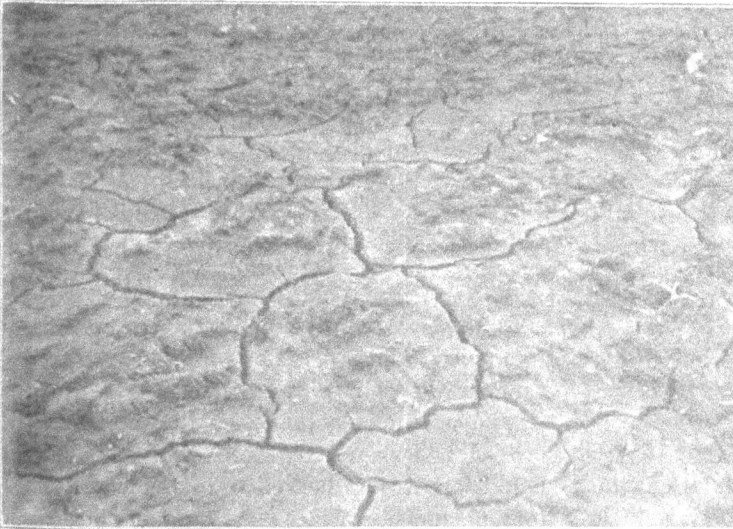

Baked Soil: Needs lime and humus, manure and compost to make
good garden soil

CULTIVATE, CULTIVATE, CULTIVATE

Cultivate Often to Kill Weeds and Save Moisture

The first requirement in gardening is proper and thorough tillage. Fall spading or plowing followed in the spring by pulverizing gives the best results. Fall plowing is best because winter freezing breaks up the soil, sods, and clods into finer particles, permits the nitrogen in the snow to soak into the soil, exposes the garden insect pests to the killing frost and birds, and permits working the soil very early in the spring.

Vacant lot in wholesale district, Chicago, before clearing off the rubbish for community garden. Cut on page 14 shows garden after planting.

USE OF SPADE OR PLOW Root crops require deep spading or plowing because their roots go deep into the soil. Leaf crops do not need deep cultivation. In spading turn the top soil so it will drop to the bottom of furrow, leave surface as even as possible and break up the soil by rapping it with the back of the spade.

USE OF RAKE AND PULVERIZER In the early spring cover the garden with manure, then spade or plow evenly, turning under the manure, and then use the garden rake or drag to pulverize the surface soil. The surface of the seed bed should be made almost level and smooth, with a slight slope from the center of the bed to the borders. Break up the lumps, bury coarse lumps and rake off coarse, large stones.

12

FINISHED   Before planting, see that soil is well pulverized
SEED BED   and the surface of the planting bed is made smooth.
        If the soil is lumpy, use a hand roller or clod
crusher.   A roller should be used on your lawn after heavy spring
rains.

MAKE YOUR   If you can't afford to buy a roller, make your
OWN ROLLER own.   Get a 24-inch glazed sewer or drain tile,
        shown as Fig. 1; carefully break or knock off
the flange (Fig. 2), set the tile on end and fasten a pipe or iron rod
into the ground in the direct center of the tile (Fig. 3).   Fill the
space round this pipe inside the tile with a mixture of concrete,
using two parts of sand, one part cement, and one part of stone
or gravel.   Fasten a steering pole or draw-bar to the pipe running through the center of the tile and you will have a service-
able roller (Fig. 4).

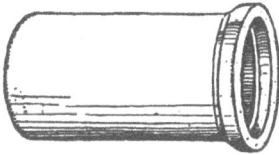

Fig. 1—Sewer Pipe

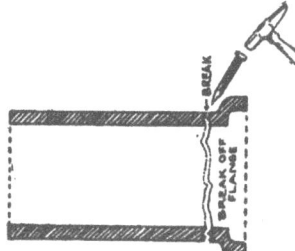

Fig. 2—Breaking off Flange

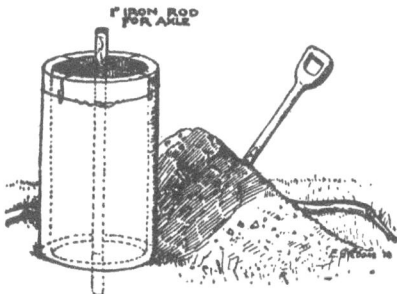

Fig. 3—Tile on end ready to be
filled with Concrete

Fig. 4—Home-made Roller
Completed

SPECIAL   Such crops as beets, onions, and radishes are
SEED BEDS grown best in beds from 4 to 8 feet wide and
       slightly raised toward center of bed.
    Plants like the cabbage are grown best on slight ridges.
Others like celery are grown in furrows or shallow trenches.

GENERAL   For general planting have the surface of the bed
SEED BEDS   made level.   When planting seeds or seedlings
use a straight edged board or garden line stretched
across the bed, and plant in perfectly straight rows.

Don't Plant Poor Seed

Do not waste vegetable seed.   They are scarce and only good seed should be planted.   Buy from reliable and established seed companies.   Do not buy more than you need.   Test your seed before planting; it will save you time and money.   Seeds are good when from 80 to 90 seeds out of every hundred will germinate.   Take your sample, count out 20 or more and test them.   To test, use a kitchen plate, cut two pieces of cloth or blotting paper, lay blotting paper in plate, wet thoroughly, drain off the surplus water, scatter the seeds evenly over blotter and cover the seeds with another blotter. Cover p l a t e with glass or another plate to prevent evaporation, and put in warm

Device for testing seed

room.   If the seeds are good they will sprout in three or four days. Keep the blotters moist.

Same tract as that shown on page 12 after it was prepared for planting.

GET AN EARLY START

Grow Seedling Plants in Your Home—Transplant in the Garden—Simple, Inexpensive—Anybody Can Do It—Will Give You Early Vegetables.

Get an old pan, washbasin, or shallow box, clean it out, and punch a few holes through the bottom to allow easy drainage (Figs. 1 and 2).

PREPARING PAN

Cover the bottom with about three inches of broken crockery, glass, or small stones. (Fig. 2.) This rough filling allows air to come up through soil and the excess water to drain out. Over this rough filling place

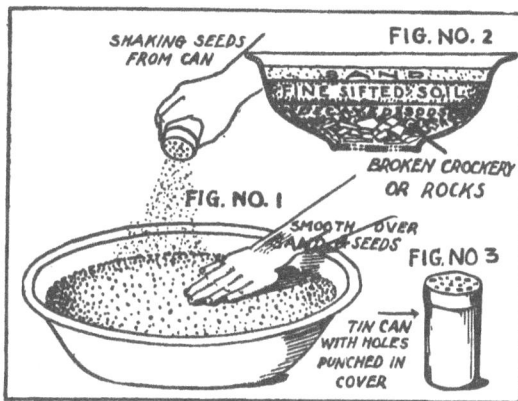

How to plant seed for germination indoors, by use of an old dishpan

about three inches of coarse soil, rotted pieces of sod, fibrous loam or leaf mold (Fig. 2), with a top layer of fine, sifted soil, sprinkled over with clean, sharp sand. If sand cannot be had use fine, sifted coal ashes.

WETTING SOIL When you have completed this filling, dip the pan or box into a tub of water so as to moisten soil before planting the seed. It is best to apply all moisture in this way. If water is carelessly applied with a sprinkling can, or tumbler, you are apt to disturb the seeds and delay their growth.

SOW SEED Distribute the seeds evenly over the entire surface of the pan. A good way to distribute the seeds evenly is to mix them with an equal amount of sand and put this mixture into a tin pepper box, (Fig. 3); any box will do. Punch holes in the cover, and sprinkle the mixture over the surface. Over these seeds sprinkle a covering of sand as thick as the seed sown.

15

HAVE A HOT BED

Advantages of a Hot Bed—Early Vegetables—But a Little Time and Labor Required—Best Assurance of a Successful Garden.

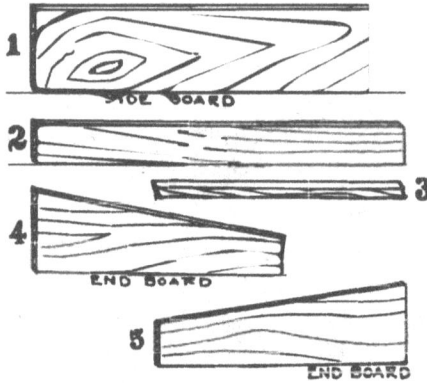

Fig. 1—The boards for hot bed

PURPOSE Plants can be started in a hotbed where the garden is permanent and of considerable size.

HOW TO MAKE A HOT BED

To make a small hotbed for home garden use, take—one board 6 feet long, 16 inches wide—(No. 1, Fig. 1). One board 6 feet long, 8 inches wide—(No. 2, Fig. 1). Two boards, 4 feet long, 8 inches wide at one end; 16 inches wide at the other end—(Nos. 4 and 5, Fig. 1).

Now nail the ends of the boards No. 1 and No. 2 to the ends of equal width of boards Nos. 4 and 5, and nail sill No. 3 across the center (Fig. 2).

As a cover for this frame use two window frames, 3 feet wide and 4 feet long. These are laid over the top to hold the heat as shown in Fig. 3. The heat is supplied by using fresh manure and providing a condition that will make it ferment.

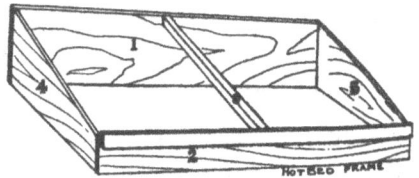

Fig. 2—Boards nailed together

Dig a pit, 6 feet long by 4 feet wide by 1 foot deep. Fill this pit with fresh manure. Add enough to bring the top 8 or 10 inches above the ground (Fig. 4).

Fig. 3—Finished hotbed box with sash

16

The hotbed frame is then set on top of this heap and the manure banked around the bottom of the frame to keep out the cold, as shown in Fig. 3 and Fig. 4. The plan and the size of the hotbed may be modified to suit conditions.

Fig. 4—Cross-section showing manure pit

SEED BED Seeds may be planted in soil spread over the manure in the hotbed, but better results are obtained by starting the seeds in what are called propagating flats.

Make and plant this flat (Fig. 1), in-doors and place it in the hotbed when planted. In filling the flat, place about two inches of rough stone and broken crockery or glass in the bottom, then about two inches of good, rich, sifted soil, run through a soil sieve—(Fig. 2), and then a top layer of clean fine sand.

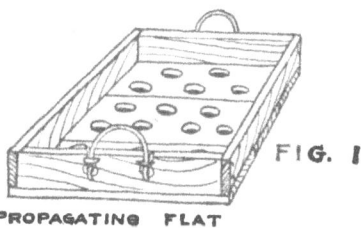

FIG. I

PROPAGATING FLAT

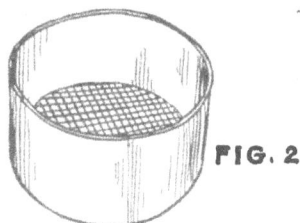

FIG. 2

SOIL SIEVE

PLANT IN FLAT When this has been done, plant your seed in straight rows, with a label naming the variety at the end of each row (Fig. 3). Water carefully so as not to disturb the seeds.

After the seeds have been properly planted, place the flat inside the hotbed, and cover with a glass window sash. The glass sash should then be covered with a canvas carpet or boards so as to keep the seeds dark while germinating. This covering can be removed after

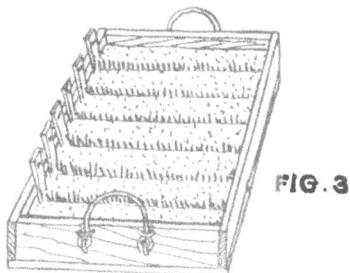

FIG. 3

GROWING SEEDS

Propagating flat

six or eight days. Keep the temperature about 70 degrees F. When the seedlings have developed two or more leaves and are crowding one another they may be transplanted into another flat, or, if the weather is favorable, into the garden.

Boys making hotbeds

Plants which grow fast, such as cabbage, cauliflower, and lettuce will get a better start by transplanting into second flat before planting into garden.

Egg plants, peppers, and tomatoes also do better after transplanting. These tender plants must be handled very carefully.

Do not let your seeds get dry, apply water carefully without force so as not to wash out the seeds.

Ventilate the hotbed by raising the sash, slightly at first, during the warm part of the day. This ventilation can gradually be increased as the seedling plants grow. This increased ventilation will "harden off" or accustom the plants to the colder temperature of the garden and help to overcome the danger of freezing.

An unsightly, unhealthy, unproductive yard (See opposite page).

TRANSPLANTING TO GARDEN

Do Not Injure Roots or Sprouts—"Water" Before Transplanting—Have Good Seed Bed.

Care must be exercised in handling the plants. Do not break or tear plants grown for their roots. Melons, cucumbers, and beans do not transplant readily. A common practice is to start the seeds of these vegetables in berry boxes, place the boxes in the hot-bed to germinate, then transplant box and all to the garden.

All seedling plants should be given a good watering a few hours before transplanting. This prevents wilting and keeps fine soil sticking to the rootlets.

Have your plant bed properly pulverized, and the surface smooth. Carry your propagating flat full of seedlings into the garden and place it beside the vegetable bed. Stretch your garden line across the planting bed, and lay your board across with the edge nearly up to the line. With the dibble, poke a hole directly on the line, being careful to make the hole large enough, place the roots of the seedling in the hole, press the soil in lightly from all sides, and the seedling is planted. The planter should rest on top of the board so as to keep surface smooth. Next it should be watered. This is best done with a fine spray, or a sprinkling can. Do not apply the water in such quantities and force as to wash the seedlings or seeds out of the soil.

Same lot shown on page 18 after cultivating. Produces vegetables for the family.

19

VEGETABLE GARDEN CHART

Kind of Vegetable	Seeds or Plants Required for 100 Feet of Row	Time of Planting out of Doors (N = North, S = South)	Distance for Rows to Stand			Depth of Planting	Time Required to Secure Crop After Planting	Approximate Yield per 100-Foot Row
			Rows Apart		Plants Apart in Rows			
			Horse Cultivation	Hand Cultivation				
Artichoke, Globe	1 oz.	Early Spring	4 to 5 ft.	2 to 3 ft.	2 to 3 ft.	1 to 1½ in.	15 mo.	400 to 500 lb.
Artichoke, Jerusalem	2 qt. tubers	Early Spring	3 to 3½ ft.	2 ft.	15 to 18 in.	4 in.	Spring to fall	4 to 10 bu.
Asparagus (for plant production)	2 oz.	Early Spring	2½ ft.	15 to 18 in.	2 to 3 in.	1 to 1½ in.	3 to 4 yr.	400 to 500 lb.
Asparagus, plants	80 to 80 plants	Early Spring	4 ft.	3 ft.	2 ft.	5 to 6 in.	1 to 3 yr.	400 to 500 lb.
Beans, bush (kidney and lima)	1 qt.	N—April to July Feb. to Apr. (Aug.)	2½ to 3 ft.	2 ft.	2 to 4 in.	1 to 2 in.	40 to 65 da.	15 to 30 qt.
Beans, pole (kidney and lima)	4 oz.	May and June	3 to 4 ft.	3 to 4 ft.	18 to 24 in.	1 to 2 in.	50 to 80 da.	10 qt.
Beets	1 oz.	Early spring	2 to 3 ft.	12 to 18 in.	4 to 6 in.	½ in.	60 to 75 da.	2 to 3 bu.
Broccoli	¼ oz.	N—May, transplant in 6 wk. May 15 to Sept.	2½ ft.	2 to 2½ ft.	1½ ft.	½ in.	100 to 130 da.	40 heads
Brussels sprouts	¼ oz.	North May to June South Jan. to July	3 ft.	2½ to 3 ft.	1½ to 2½ ft.	½ in.	100 to 125 da.	30 qt.
Cabbage, early	½ oz.	N—Mar. and April (start in hotbed during Feb.) Oct. to Dec. S—	2½ to 3 ft.	2 to 2½ ft.	14 to 18 in.	½ in.	110 da. from plants	40 heads
Cabbage, late	½ oz.	N—May and June June and July S—	30 to 40 in.	24 to 36 in.	18 to 24 in.	½ in.	150 da. from plants	40 heads
Carrot	1 oz.	Early spring, May 15 to June 15	20 to 30 in.	12 to 18 in.	3 to 6 in.	¼ to ½ in.	80 to 110 da.	20 bunches

Reprinted by the courtesy of Ralph E. Weeks, President of the International Correspondence Schools, Scranton, Pennsylvania. (See Note Page 25). Copyright 1917 by International Textbook Co.

Kind of Vegetable	Seeds or Plants Required for 100 Feet of Row	Time of Planting out of Doors (N = North, S = South)	Distance for Plants to Stand: Rows Apart, Horse Cultivation	Distance for Plants to Stand: Rows Apart, Hand Cultivation	Distance for Plants to Stand: Plants Apart in Rows	Depth of Planting	Time Required to Secure Crop After Planting	Approximate Yield per 100-Foot Row
Cauliflower	¼ oz.	May 15 to June 15 (start in hotbed in March)	2½ ft.	2 to 2½ ft.	1½ ft.	½ in.	100 to 130 ds.	40 heads
Celeriac	⅛ to ¼ oz.	May and June (start in cold frame in April)	2½ to 3 ft.	1½ to 2 ft.	6 to 8 in.	⅛ in.	100 to 150 ds.	150 to 200 plants
Celery	¼ oz.	May for early crop (start under glass in Mar. or April) July for late crop (sow seed in May)	3 to 6 ft.	1½ to 3 ft.	4 to 8 in.	⅛ in.	160 to 170 ds. from seed 90 to 100 ds. from plants	150 to 300 heads
Chard (Swiss)	2 oz.	Early spring to June 15	2 to 3 ft.	1½ to 2 ft.	4 to 6 in.	½ in.	To middle of summer	150 to 200 lb.
Chicory	1 oz.	N— May and June S— Mar. and April	2½ to 3 ft.	1½ to 2 ft.	4 to 5 in.	½ in.	150 ds.	100 to 150 lb.
Chinese cabbage	¼ oz.	Early spring or July and Aug.	2 ft.	20 in.	15 in.	½ in.	120 to 130 ds.	70 to 80 heads
Chives	120 clumps	N— Early spring S— Feb. to May	2 to 2½ ft.	18 to 20 in.	10 in.	½ in.	60 to 90 ds.	200 clumps
Citron (watermelon type)	1 oz.	N— May 15 to June 15 S— Mar. to May	5 to 6 ft.	5 to 6 ft.	5 to 6 ft.	¾ to 1 in.	100 to 120 ds.	50 fruits
Corn salad	2 oz.	N— Early spring to Sept. S— Aug. to frost	2 to 2½ ft.	12 to 18 in.	6 in.	½ to 1 in.	60 to 65 ds.	200 heads
Corn, sweet	2 to 4 oz.	Early spring	2½ to 4 ft.	2 to 3 ft.	12 to 18 in.	½ to 1 in.	60 to 100 ds.	12 doz.
Cress, upland	¼ oz.	N— Late spring S— Late fall or spring	2½ ft.	12 to 18 in.	4 to 6 in.	¼ in.	7 to 8 wk.	150 bunches
Cress, water	¼ oz.	N— April to Sept. S— Early spring		18 in.	6 to 8 in.	In 1½ to 3 in. of water	6 to 8 wk.	200 bunches
Cucumber	½ oz.	N— May to June S— Mar. to April	4 to 6 ft.	4 to 5 ft.	4 to 5 ft.	½ to 2 in.	60 to 80 ds.	10 doz.
Dandelion	½ oz.	April to June	2 to 2½ ft.	12 to 14 in.	3 to 4 in.	¼ to ½ in.	6 mo. to 1 yr.	3 bu.

Kind of Vegetable	Seed or Plants Required for 100 Feet of Row	Time of Planting out of Doors (N=North, S=South)	Distance for Plants to Stand			Depth of Planting	Time Required to Secure Crop After Planting	Approximate Yield per 100-Foot Row
			Rows Apart		Plants Apart in Rows			
			Horse Cultivation	Hand Cultivation				
Dill	½ oz.	Early spring or late summer	20 to 24 in.	16 to 18 in.	8 in.	½ in.	Spring to fall	20 bunches
Eggplant	½ oz.	N— June, S— Jan. to May	3 to 5 ft.	3 to 4 ft.	2 to 3 ft.	½ to 1 in.	120 to 150 da.	130 eggs
Endive	1 oz.	N— July or Aug., S— Aug. or Sept.	2 to 2½ ft.	12 to 20 in.	12 in.	½ in.	8 to 12 wk.	80 to 100 plants
Fennel	½ oz.	Intervals of 6 to 8 wk. Very early spring	2 ft.	15 to 24 in.	8 to 10 in.	½ in.	120 to 150 da.	100 bunches
Garlic	300 to 400 cloves	N— Early spring, S— Oct. to Mar.	2 to 2½ ft.	12 to 18 in.	4 to 6 in.	1 to 2 in.	4 to 5 mo.	75 bunches
Horseradish	65 to 80 roots	Early spring	2 to 3 ft.	2 ft.	15 to 18 in.	Top 1 to 2 in. below surface	1 to 2 yr.	40 to 50 lb.
Kale, or borecole	½ to 1 oz.	N— May to June, S— Aug. to Sept.	2½ to 3 ft.	18 to 24 in.	8 to 12 in.	½ in.	Ready for use after frost	40 plants
Kohlrabi	½ oz.	N— April to June, S— Sept. to Mar.	2 to 3 ft.	15 to 24 in.	6 to 8 in.	½ in.	75 to 120 da.	175 plants
Leek	1 oz.	N— July (sow seed under glass in April or May), S— Feb.	2 to 3 ft.	12 to 20 in.	5 in.	1 in.	120 to 180 da.	200 to 240 plants
Lettuce	½ to 1 oz.	N— April to Aug., S— Oct. to Mar.	2½ ft.	12 to 18 in.	12 in.	⅛ to ¼ in.	60 to 90 da.	100 heads
Melon, muskmelon	½ oz.	N— May 15 to June 15 (sow seed under glass April 15), S— Feb.	6 ft.	Hills 6 ft.	4 ft.	1 to 2 in.	110 to 130 da.	80 melons
Melon, watermelon	1 oz.	N— May 15 to June 15, S— Mar. to May	7 to 10 ft.	7 to 10 ft.	7 to 10 ft.	½ to 1 in.	115 da.	50 melons

Kind of Vegetable	Seeds or Plants Required for 100 Feet of Row	Time of Planting out of Doors N = North S = South	Distance for Plants to Stand: Rows Apart, Horse Cultivation	Distance for Plants to Stand: Rows Apart, Hand Cultivation	Distance for Plants to Stand: Plants Apart in Rows	Depth of Planting	Time Required to Secure Crop After Planting	Approximate Yield per 100-Foot Row
Mustard	½ to 1 oz.	Mar. to May (Sept.)	2 to 3 ft.	12 to 15 in.	4 to 6 in.	½ in.	60 to 90 da.	200 heads
New Zealand spinach	1 oz.	Early spring	3 ft.	24 to 36 in.	12 to 18 in.	1 to 2 in.	60 to 100 da.	100 lb.
Okra, or gumbo	1½ to 2 oz.	N— May to June S— Feb. to April	4 to 5 ft.	3 to 4 ft.	14 to 36 in.	1 to 2 in.	115 da.	1,000 pods
Onion, seed	½ oz.	N— April and May S— Oct. to Mar.	2 to 2½ ft.	12 to 18 in.	2 to 4 in.	¼ in.	135 da.	2 bu.
Onion, sets	1 to 1½ qt.	N— Early spring S— Fall and Feb.	2 to 3 ft.	12 to 18 in.	2 to 4 in.	½ in.	60 da.	75 bunches
Parsley	¼ oz.	N— Last of Mar. to 1st of April S— Sept. to May	2 to 2½ ft.	12 in.	3 to 6 in.	½ in.	95 to 120 da.	200 plants
Parsnips	½ oz.	April and May	20 to 30 in.	12 to 15 in.	3 to 4 in.	½ in.	125 to 160 da.	3 bu.
Peas	1 qt.	N— Early spring S— Dec. to April	2½ to 4 ft.	2 to 3 ft.	10 to 15 in.	1 to 2 in.	60 to 80 da.	4 pk.
Pepper	¼ oz.	N— May and June (start early plants under glass in Mar.) S— Last of Mar.	2½ to 3 ft.	20 to 24 in.	18 to 24 in.	½ to 1 in.	100 to 140 da.	75 doz. pods
Peppermint	100 root clumps	Early spring	2½ ft.	20 to 24 in.	12 in.	1 to 2 in.	120 da.	100 bunches
Potato, sweet	5 lb. or 75 slips	May & June (start in hotbed in April)	3 to 4 ft.	2½ to 3½ ft.	14 to 18 in.	3 in.	140 to 160 da.	2 bu.
Potato, white	5 to 8 lb. tubers	N— Mar. to June S— Jan. to April	2½ to 3 ft.	24 to 36 in.	12 to 18 in.	4 in.	80 to 140 da.	2 to 3 bu.
Pumpkin	1 oz.	N— May to June S— April and May	8 to 10 ft.	8 to 10 ft.	2 plants in hill 8 to 10 ft.	1 to 2 in.	110 to 140 da.	30 pumpkins
Radish	1 oz.	N— Mar. to Aug. S— Sept. to April	2 to 2½ ft.	6 to 12 in.	1 to 2 in.	½ in.	30 to 40 da.	75 bunches

Kind of Vegetable	Seeds or Plants Required for 100 Feet of Row	Time of Planting out of Doors (N = North, S = South)	Distance for Plants to Stand: Rows Apart, Horse Cultivation	Distance for Plants to Stand: Rows Apart, Hand Cultivation	Distance for Plants to Stand: Plants Apart in Rows	Depth of Planting	Time Required to Secure Crop After Planting	Approximate Yield per 100-Foot Row
Rhubarb seed for plant production	1 to 3 oz.	N— Early spring S— Fall or spring	4 to 4½ ft.	3 to 4 ft.	2 to 4 ft.	½ in.	2 to 4 yr.	100 bunches
Rhubarb, plants	25 to 50 roots	N— Early spring S— Fall	4 to 4½ ft.	3 to 4 ft.	2 to 4 ft.	2 to 3 in.	1 to 3 yr.	100 bunches
Rutabaga	½ to 1 oz.	N— June to July S— Aug. to Sept.	2½ to 3 ft.	18 to 24 in.	8 in.	½ to ¾ in.	70 to 100 da.	2 bu.
Sage	1 oz.	N— June 15 to July 1 (plants) S— Fall or early spring	20 to 24 in.	15 in.	10 in.	¼ to ½ in.	140 to 160 da.	100 bunches
Salsify	1 oz.	Early spring	2 to 2½ ft.	12 to 15 in.	3 in.	½ to 1 in.	130 to 160 da.	35 bunches
Savory	1 oz.	N— April or May S— Fall or early spring	20 to 24 in.	12 to 18 in.	Winter 2 in. Summer 6 in.	¼ to ½ in.	140 to 160 da.	100 bunches
Sea Kale	2 oz.	Early spring	2 to 3 ft.	2 ft.	18 to 24 in.	1 in.	2 yr.	100 lb.
Spearmint	100 root clumps	Early spring	2½ ft.	2 ft.	12 in.	½ in.	120 da.	100 bunches
Spinach	1 oz.	N— April 1 to Sept. 15 S— Aug. to Oct.	2 to 2½ ft.	12 to 18 in.	6 to 8 in.	¼ in.	50 to 60 da.	1 bu.
Squash, bush or early	1 oz.	N— May S— Feb. to Mar.	3½ to 4 ft.	3½ to 4 ft.	3½ to 4 ft.	1 in.	60 to 75 da.	250 squashes
Squash, late	1 oz.	N— May S— Feb. to Mar.	6 to 10 ft.	6 to 10 ft.	6 to 10 ft.	1 in.	125 to 150 da.	40 squashes
Sweet basil	1 oz.	April	2½ to 3 ft.	2 ft.	6 in.	¼ to ½ in.	120 to 150 da.	100 bunches
Thyme	1 oz.	April	2½ to 3 ft.	18 to 20 in.	6 in.	¼ to ½ in.	120 to 150 da.	100 bunches
Tomato	¼ oz.	N— May to June S— Jan.	3 to 5 ft.	3 to 4 ft.	2½ to 4 ft.	½ in.	130 to 150 da.	12 bu.
Turnip	½ to 1 oz.	Early spring	2 to 2½ ft.	12 to 18 in.	2½ to 8 in.	½ to ¾ in.	40 to 60 da.	2 bu.

NOTES FOR VEGETABLE GARDEN CHART.—Dates in the foregoing table indicate the time a second crop may be planted the same season. The dates of planting are approximate and will vary with the season. The dividing line between North and South is considered to be the continuation of the southern boundary of Pennsylvania or about the 40th parallel of latitude. Unless you have a large plot do not attempt to grow all the vegetables in this table. Usually the small home gardener will find it more advisable to buy plants of such vegetables as asparagus (preferably 1-year-old roots), cabbage, cauliflower, celery, eggplant, pepper, tomato, etc., than to grow them himself.

TIME
TO
PLANT
SEEDS
No definite rule can be given for planting the seeds out of doors, except in a general way, as shown in the table beginning on page 20. The time for planting varies with the location, the character of the soil, and weather conditions. Wait until the frost is out of the ground and the season is far enough along to assure warm weather.

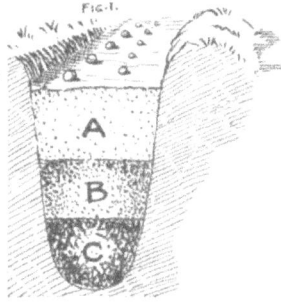

Method of sowing peas: A.—Fine soil. B.—Coarse soil. C.—Mixture of manure and soil

PLANT
PROTECTION
Young plants sometimes need to be protected against intensive heat, freezing and wind.

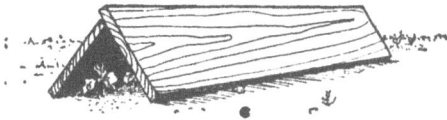

Fig. 1—Winter protection for plants

This protection is provided by setting up two boards to form an inverted "V" running lengthwise with the plant rows. (Fig. 1.) A covering of loose straw is a good protection against frost. For a few plants an upturned glass jar or tumbler will do. Pine boughs, leaves and straw are often used for winter protection.

REPEAT
PLANTING
Radishes, lettuce, carrots, cabbage and corn should be planted at intervals so as to provide a continuous supply throughout the season. Early and late varieties of corn, peas, beans, etc., should be planted. Extra early sorts may be followed by early, medium and late varieties, thus giving a constant supply of good vegetables.

CARE OF THE GARDEN

**Cultivate Often—Keep Down the Weeds—Thin Out Largest
Plants First—Tomatoes, Lima Beans, Pole Beans,
Egg Plant, Peas, etc., Should be Fastened
to Stakes—Combat Insects and
Plant Diseases.**

Cultivate often; it conserves the moisture for dry weather. Keep a "dust mulch" on the surface. A fine pulverized surface layer of soil prevents the moisture from escaping by evaporation.

SUGAR EXPERIMENT You can show just how moisture comes up in the soil, and how the soil mulch keeps it from evaporating. Take a cube of ordinary loaf sugar. On top of it heap powdered sugar about a quarter of an inch deep. (See Fig. 1, following page.) The cube of sugar represents the soil, and the powdered sugar the dust mulch you make by stirring the top soil.

Now take a butter pat or a tumbler with a concave bottom. Put a few drops of ink in water to color it. Put this water into the dish. Place the cube of sugar into this (See Fig. 2). See how quickly the water rises to the top of the cube. In 10 seconds it reaches the top of the loaf sugar but you will have to let it stand an

Boys can grow back yard gardens and furnish vegetables for the family

26

hour or more before the water soaks through the powdered sugar. In just this way a dust mulch keeps the water in the soil.

The finer and firmer the seed bed, the easier it is for the water to rise from below. This is why loose, coarse manure or large

Fig. 1—Experiment showing how water comes up in the soil and how a soil mulch keeps it from evaporating

Fig. 2—Water reaches the top of loaf sugar in ten seconds, but it takes an hour or more for it to soak through the powdered sugar

Loaf sugar represents the soil; powdered sugar represents the soil mulch

weeds plowed under in the spring, make a space which keeps the water down in the ground from coming up to the plants. The seed bed should be packed firm below but have a loose dust mulch two or three inches deep on top.

STIR THE TOP SOIL The ground between plants should be stirred with a garden rake, hoe, or cultivator as often as necessary to keep the top soil from getting hard or crusty and to keep down the weeds. Water when soil and plants are dry. Apply water with a garden hose or sprinkling can, and try to imitate rain as nearly as possible. That is, apply the water in a fine spray rather than a heavy stream. Give plants a heavy application at each watering; light sprinkling that merely wets the surface is one of the surest ways of spoiling your plants, because the top soil bakes.

KEEP DOWN   Do not let the weeds get a start in your garden.
THE WEEDS   Pull them out by the roots or dig them out
with a hoe, rake or hand weeder. Weeds
shade the young plants and use the water and plant food needed
by the vegetables. Never let weeds go to seed.

THINNING   Thin and cut your largest plants first. In this
way you thin out the rows, allowing the remaining
plants to develop better vegetables. Lettuce, onions, carrots,
turnips, radishes, etc., should be thinned out along the rows.

Pull the largest vegetables and leave small ones to grow

CLEANLINESS   Whenever you harvest vegetables be sure
to remove any refuse that remains. Do not
let the leaves of plants lie in the garden. They attract flies and
make paths soft and unsightly. Throw all the refuse on the
compost pile where it will decay and make fertilizer for next
year. Many garden insects live through the winter under this
loose material left in the garden. Dead vines, leaves and roots
are sometimes covered with disease spores, insect eggs and pupae,
and should in that case be burned.

STAKING Plants such as tomatoes, lima and pole beans, eggplant, peas, cucumbers, and celery are fastened to stakes, trained on supports, tied up, etc. This is done to keep the vegetables off the ground, to keep the plants from breaking, and to give better exposure to the sun. Use contrivances shown in illustrations.

Support for cucumbers
or peas

Supporting pole beans
and peas with sticks

Support for toma-
toes or peas

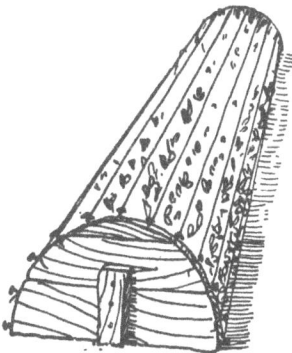

Support for green beans,
peas and cucumbers

Protectors and supports
for beans and peas.
Tree twigs

KILL INSECT PESTS

Insect Pests Damaging Vegetable Crops are of Two Classes: Chewing and Sucking Insects. Chewing Insects Eat the Plants—Sucking Insects Suck the Juice with Their Tube-like Beaks.

Chewing insects are destroyed by spraying the plants upon which they feed with an arsenical poison, Paris green or arsenate of lead. Of these, arsenate of lead is the best and more extensively used. Arsenate may be obtained in a paste or powdered form. If you use large quantities get the paste form, but where small quantities are used the powdered form is best. The paste hardens and deteriorates while the powder keeps its strength. The powdered form may be dusted upon the plant, or, when a spray pump is available, made into a paste and then diluted with water. A large tablespoonful is enough for a gallon of water. If it is strong enough it should leave a white film on the leaf when dry. All chewing insects can be killed with this poison.

The cut worms which chew on the stems and roots of vegetables at the surface of the ground, are killed by scattering a bran mash poisoned with arsenate of lead on the ground around the plant. The cut worm will eat this poisoned bran.

Cabbage, cauliflower, etc., should not be sprayed with arsenate just before they are to be eaten. In such cases hellebore can be dusted upon the plant, or use a tablespoonful in a quart of water for spraying.

Chewing Insects—Cut-worms eating leaf of cabbage plant. Kill by spraying early with arsenate of lead

30

HOW TO COMBAT SUCKING INSECTS Sucking insects such as the Aphis, plant lice, scale insects, mealy bugs, mites, red spiders, etc., are killed by spraying the insect and plant with kerosene emulsion or some other caustic solution. Fine dust sprinkled generously on plants attacked by the Aphis will kill the insects by smothering them.

HOW TO MAKE KEROSENE EMULSION Dissolve half a bar of soap in two quarts of hot water, add one gallon of kerosene and then churn or agitate vigorously, until it is thoroughly mixed and no globules of oil can be found floating upon the surface. This makes a stock solution which should be diluted before using by adding for summer spraying about 12 to 15, and for winter spraying, from 8 to 10 times as much water.

Kill root maggots by soaking the ground around the plant with kerosene emulsion. Tobacco stems or dust scattered over the surface of the ground around plant will keep off insects and act as fertilizer. Pyrethrum powder when dusted upon the plant will kill the sucking insects.

Keep these poison spray mixtures out of the reach of children and careless persons. Label materials POISON in big letters.

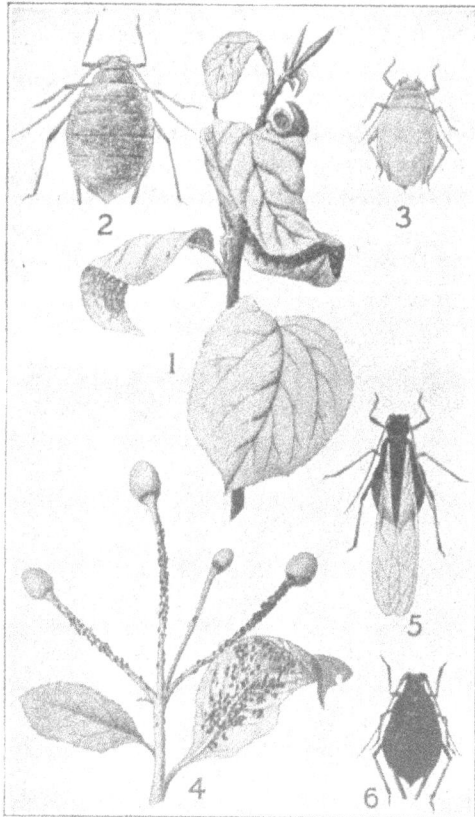

Aphis or Plant Lice—Most destructive sucking insects to vegetable crops. Figs. 2 and 3—Young Aphis. Fig. 5—Male adult. Fig. 6—Female adult. Fig. 4—Young Aphis on plant.

How to Prevent Fungus Diseases

All plant diseases work practically the same way. They rot, or transform the plant tissues. It is easier to prevent them from getting a start upon the plant than to check them after they have started. We spray as a preventive rather than as a cure.

USE BORDEAUX MIXTURE FOR SPRAYING Bordeaux mixture is the best spray for plant diseases. Make your own Bordeaux mixture. Take a fruit jar, crock or wooden pail. Put two quarts of water in this vessel. Put half a pound of copper sulphate in a piece of cloth and hang it to a stick laid

Potato Blight. A fungus disease of potato

Spraying with bucket pump

across the top of pail or crock suspended in the water. In another pail put one-half a pound of quick lime, add warm water to dissolve, making a thick creamy solution. Strain each dissolved solution through a cloth, and pour the two solutions into a third vessel at the same moment, so they will come together and mix at the bottom of the vessel.

For Blight on potatoes use a spray made of 4 pounds of lime, 4 pounds of copper sulphate to 50 gallons of water.

HOW TO Apply
SPRAY Bordeaux
mixture
on a calm day so it
will not be blown
away by the wind.
Use a pump with

Fig. 1—Cheap spray pump, made of tin or brass. Throws a spray as fine as mist. Used to spray animals and crops. Best cheap sprayer on the market

plenty of force and a nozzle that will throw a fine spray. Spray all parts of the plant so that a fine film of the spraying solution used will cover the surface of the plant tissues.

Potatoes should be sprayed with Bordeaux when they are about 6 inches high, this is done to prevent Potato Blight from attacking the plant.

Navy Beans and Tomatoes should also be sprayed early with Bordeaux as a precaution or preventive against blight and wilt.

HAVE A A small hand bucket
SPRAY spray pump (Fig. 2),
PUMP is very good for spraying, and for other
general uses such as spraying the chicken coop or whitewashing the cellar. A tin pump, as shown in Fig. 1, will do the work in the small garden. I have seen a whisk broom used, the liquid spray being applied by soaking it in the liquid and then switching the liquid upon the plant.

DUST Powders are dusted up-
SPRAYS on the plants or blown
against them by means
of small atomizers or dust blowers. These sprays will smother the insects by stopping up the breathing pores.

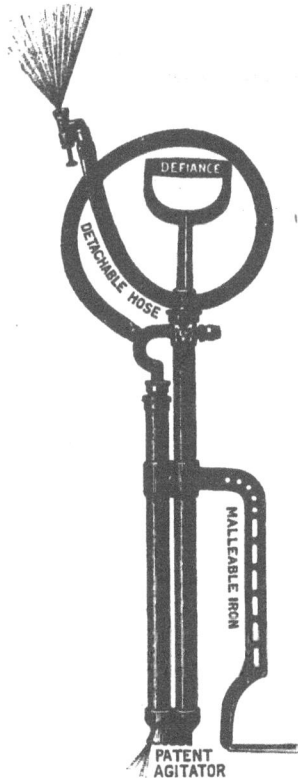

Fig. 2—Hand Bucket Spray Pump. Best for Home and Garden Use

SAVE—DON'T WASTE—SAVE

Whenever You Have a Surplus of Vegetables, Can, Dry or Store Them. Do Not Let Them go to Waste.

Store your surplus of root crops: Potatoes, carrots, winter radishes, turnips, parsnips, beets, cabbage, and celery. You can store in pits, that is, bury them in trenches under one foot of soil, or bury in barrels or boxes with about one foot of soil over top, as seen in the illustrations.

In any case do not cover too closely until it becomes freezing cold. Some slight ventilation should be provided. This can be supplied by having a pipe or tile run upright through the pile of stored vegetables, and through the dirt covering. This pipe should be covered with a hood to keep the rain out, and stuffed with straw during severe freezing weather. When vegetables are needed

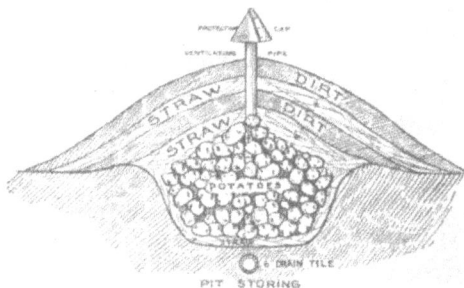

Pit Storage: Irish potatoes stored out of doors. Make pit in well-drained location. Have layer of straw 3 inches thick. Cover potatoes with 3 inches of straw

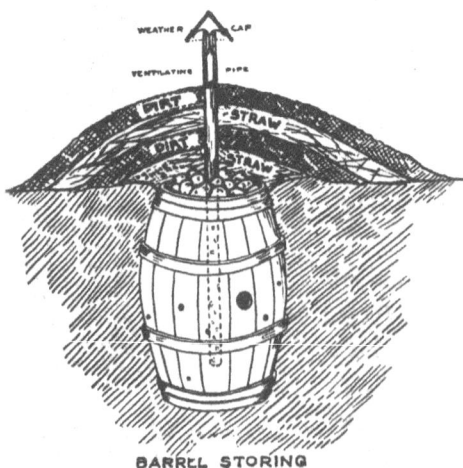

BARREL STORING

take them out during the warm part of the day so there will be less danger of freezing.

Only those people who store vegetables for winter use can appreciate the saving that can be made in this way. If at any time during the winter the stored vegetables begin to decay, do not lose them —they can be dried.

34

A Few of the Best Vegetables

CHILDREN'S HOME AND SCHOOL GARDENS On city vacant lots, children's gardens, and in poor soils do not try to grow rare or uncommon varieties of vegetables. The beginner should grow those varieties that are sure, need the least attention, and are least affected by insects and diseases especially. Grow such as beans, peas, tomatoes, beets, sweet corn, radishes, carrots, Swiss chard, lettuce, onions, cabbage, cucumbers, squash and rhubarb.

These give the best results and respond best to the efforts of the child.

The cauliflower, artichoke, asparagus, Brussels sprouts, eggplant, etc., can be added where the soil is good and work is carried on by an experienced gardener.

Who Can Grow A Garden

SCHOOL CHILDREN School children have proven themselves worthy tillers of the soil. Their ability to grow valuable crops has been recognized by the Government.

MAKE GARDENING A REGULAR SCHOOL STUDY School gardening work has become a part of the regular school work in many of our country and city schools. We need more competent teachers and practical directors for this gardening work.

A useful recreation. A producer reduces family expenses

GOVERNMENT SUPPORT The Government has employed one of our ablest educators to supervise and direct the work of organizing Boys' and Girls' Club work. Their principal object is to produce and conserve food.

TEACH IN TERMS OF THE CHILD School gardening must be considered in its relation to the child's daily life, its effect upon the character of the child, its place in the curriculum, and its relation to the other subjects in the course of study.

GET EFFICIENT LEADERS In most cases the older boys and girls can be organized into garden clubs by the teachers and the work of these clubs can be supervised and directed by some club worker, Boy Scout Master, Y. M. C. A. leader, or Mother's Club.

Gardens on school grounds should be under the supervision of the principal or teacher.

School gardens as a part of the regular school work furnish a source of:

1. Nature study material.
2. Art work—design, color, form, grouping, etc.
3. Language—topics, composition, literature.
4. Mathematics—measuring, plotting, figuring seeds and plants needed.
5. Physics and Chemistry—natural forces.
6. Domestic Science—raw food stuffs, cooking, canning, drying, storing necessities of the home.

Oakland School Garden Lot—Unsightly vacant lot at Cottage Grove and Bowen Avenues, Chicago, before improving with Children's Garden. Lot 100 x 100 feet square

Cottage Grove Ave.

School Children's Garden.
Oakland School
Chicago Ill.

Oakland School lot at Cottage Grove and Bowen Avenues after
establishing a Children's School Garden. Children doing
gardening. Parents' Club directing the work.

7. History—plants, old customs, trade routes, industries.

8. Training in industrial work—food production.

9. Manual training—making tools and accessories.

10. The child becomes a factor in the community.

11. He can earn money to help pay the expenses of his education.

Get children interested in growing things at home, let the children have a part of the back yard garden to work as their own.

FACTORY GARDENS Factory employers are encouraging the growing of vegetables by their employees. It helps reduce their living expense and promotes a healthy co-operative spirit.

GARDEN PROVIDED BY EMPLOYERS The Stock Yards Packing Companies, of Chicago, have provided a fifty-acre vegetable garden for one thousand or more of their employees and people of the community for individual gardens. Each plot is twenty feet wide and one hundred twenty feet long. The Companies furnished the land, fertilizers, plowing, supervision and direction.

The employees pay a 50 cent annual fee, furnish their own tools, plant and care for the gardens and have all they grow.

These gardens are operated under the auspices of the Stock Yards Community Clearing House.

The McCormick Tractor Works and the Wisconsin Steel Mills gardens are operated on the same plan.

Factory Employes Garden—Fed 1000 families

HOME YARD GARDENS Home yard gardens mean the most for the country and family. Kitchen gardens will

build up the home and keep the entire family interested. Wherever you find a home yard where grass, flowers and vegetables have taken the place of cinders, tin cans and rubbish, there you will find a happy family, a successful father, a contented m o t h e r and well-b e h a v e d c h i l d r e n. Here in their own garden they all have a common interest. Each has his favorite flower a n d vegetable and while

WHAT 2387 COOK. COUNTY BOYS AND GIRLS DID WITH GARDENS

1916

COOK COUNTY GARDEN CLUB MEMBERS	2387
ACRES CULTIVATED	261.6
TOTAL INCOME	$47,896.53
TOTAL COST	6,267.25
PROFIT	$41,629.28

1915

CLUB MEMBERS	1778
ACRES CULTIVATED	205.3
GROSS RECEIPTS	$37,215.53
EXPENSES	5,186.28
NET RECEIPTS	$32,029.25

each contributes a part of his or her time, all receive the benefits of the whole. A child brought up in surroundings such as these is surely making better preparations for the future than those who are spending their youth in useless or idle play.

Lawn with vegetable garden in rear

MAKE POTATO WAR BREAD

100,000,000 Bushels of Potatoes Will Save 100,000,000 Bushels of Wheat—Potato Bread is Better

By P. G. Holden

It is hardly possible to estimate from a standpoint of food conservation the great value of potatoes as a substitute for wheat and other grains in the making of bread.

One hundred million bushels of small potatoes will save 100,000,000 bushels of wheat. About 30 per cent of the annual potato crop in the United States consists of small potatoes which are unmarketable, made little use of, practically wasted, almost a total loss to the country. The small potatoes can be substituted for one-third the wheat flour used in making bread. Potato bread is better bread in every way than bread made entirely of wheat or a combination of wheat and other grains.

The use of potatoes in bread is economical at any time. It is patriotic at this time; it utilizes waste potatoes; saves wheat and other grain which can be exported; saves corn, barley and oats which can be used to produce meats and fats for our soldiers; gives us white, moist and wholesome bread for every meal; does not require extra work for the housewife, nor change the usual custom and practices of the home; means two and one-third wheatless days a week, ten wheatless days a month, four wheatless months a year.

Potatoes are grown in every section of the country, found in every home, are a universal food. Every family can produce them. Potatoes are an abundant crop in the United States; the possibilities of increasing the total yield in this country cannot be estimated. Potatoes are a perishable crop—cannot be exported to foreign countries, cannot be carried over from one season to another. The potatoes we grow in this country we must use at home. The nature of potato starch is so nearly the same as wheat flour that there is no difficulty in using this combination in bread making.

Last season the farmers of the United States produced about 440,000,000 bushels of potatoes. About 100,000,000 bushels of this crop were small, irregular, unmarketable potatoes.

I earnestly believe that in no other way can so great a saving be made in food in America with so little labor and so small expense as the use of small potatoes in the making of bread in place of small grains which can be shipped to our soldiers and the fighting armies of the allies to help us win this war.

40

How to Make Potato War Bread

⅔ cup sweet milk
1 cup potato
2 cups flour
1 teaspoon salt
1 teaspoon sugar
½ yeast cake
These measurements make one loaf. Increase ingredients according to number of loaves you wish to make. One yeast cake will make 3 or 4 loaves.

Heat milk to boiling point, then cool to luke-warm. Bake or boil potatoes, then mash or put thru ricer. Dissolve yeast cake in the milk. Make a sponge as follows: mix milk, yeast cake salt, sugar, all the mashed or riced potatoes and ⅓ of the flour. Beat well, let stand overnight to rise. In the morning add balance of flour—let rise again until double in bulk, then mold into a loaf; let rise again to double in bulk, then bake 40 minutes in a moderate oven. A little more flour will be required if potatoes are not mealy.

Potato Biscuits

2 cups flour
1 cup potato
3 teaspoons baking powder
1 scant teaspoon salt
1 tablespoon butter or lard
1 teaspoon sugar
Sweet milk to make a dough which can be rolled for biscuit.

Sift flour, baking powder, salt and sugar together. Work butter or lard into flour, add potatoes which should be boiled or baked and put thru ricer, then add milk to make a dough which can be easily handled on board. Roll out about ½ inch thick, cut with biscuit cutter and bake 15 minutes in a quick oven.

Potato Doughnuts

1 cup sugar
½ teaspoon shortening
1 egg
½ cup sweet milk
¼ teaspoon cinnamon
¼ teaspoon nutmeg
2 teaspoons baking powder
1 cup riced potatoes
2 cups flour
½ teaspoon salt

Mix sugar, spices, salt and shortening. . Add well-beaten egg and milk. Beat well and add flour and baking powder which have been sifted together. Mold on board and roll to ½ inch thick, cut with doughnut cutter and fry in deep fat.

SUMMARY

1st. Unusual conditions brought on by the war make the Vegetable Garden a very important factor in food production and conservation.

2d. The Home Garden will save wheat, meats and fats, so much needed by our soldiers abroad.

3d. Plan the garden in long rows and cultivate with a horse cultivator whenever possible. On the farm this can be conveniently done. The vacant-lot garden in the city can be taken care of with hand cultivators. Gardens will clean up the weeds, increase the value of property, utilize waste lands, provide useful, healthful occupation, lessen the cost of living, provide a variety of foods.

4th. In every large garden should be planted staple products, such as potatoes, tomatoes, turnips, beets, beans, peas, etc. These are crops that can be stored, dried and canned for the winter.

5th. For the best results the garden should be manured and plowed in the fall; fertilize in the spring, make well pulverized, but firm, seed bed. Plant early and late varieties; have vegetables throughout the season. Cultivate often to save moisture and kill the weeds.

6th. If the soil is sour test it with blue litmus paper and apply lime. Sour, poor, weedy soil will not grow a garden.

7th. Plant the best seed you can buy; have it tested. Start seed indoors and in hotbeds (see directions for making hotbeds on page 16). This work can be done at home and costs very little.

8th. In transplanting seedlings handle the young plants carefully. Do not break or injure the stem or roots. Water all seedlings before resetting or transplanting.

9th. For planting instructions see chart page 20. This chart will give kinds of vegetables to grow, time of planting, amount of seed required, depth to plant, time required to secure a crop and approximate yield.

10th. In harvesting vegetables be careful to remove any refuse that remains on the plant. Do not let the vegetables lie in the garden. Keep them away from the flies. Throw the tops on compost pile for fertilizer for

42

the next year. Collect vines, leaves and roots affected with disease, insect eggs., etc., and burn them.

11th. Tomatoes, beans, peas and the trailing plants should be fastened to stakes. This is done to keep the vegetables off the ground and to give better exposure to the sun.

12th. For methods of combating insect enemies to vegetable crops and the treatment for fungus diseases see pages 30 and 31.

13th. Bordeaux Mixture is one of the best sprays for plant diseases. It can be made at home and costs very little. (See directions for making on page 32.) This mixture is used for potato blight and other fungus diseases.

14th. The Spray Pump is indispensable in raising a vegetable garden. A small hand spray pump can be purchased for a few cents.

15th. For full information on Home Canning by the Cold Pack Method write to the Agricultural Extension Department, International Harvester Company, Chicago, for 64-page booklet, giving full directions. Enclose 5 cents for postage.

16th. School gardens have become a part of the regular work in many country and city schools. In many states garden work is included in the regular school work the same as arithmetic, language and other lessons. School gardens should be considered in relation to the child's daily life. (For particulars see page 36.)

17th. In nearly every large city, the manufacturing districts, companies have found it good business to encourage the growing of vegetables by their employees. The vegetable garden helps to reduce their living expense and helps to promote a healthy co-operative spirit.

There should be a garden for every family in America, not only in the country, but in the cities as well. Farmers should prepare to take care of the surplus of their gardens by having root houses for the storage of potatoes and other root crops. This is a patriotic duty as well as good business now in this time of war.

THE FOLLOWING IS FOR HISTORICAL PURPOSES ONLY
AND DOES NOT REFLECT CURRENT
SCIENTIFIC KNOWLEDGE, POLICIES,
PRACTICES, METHODS OR USES.

Home Drying

of

Fruits

and

Vegetables

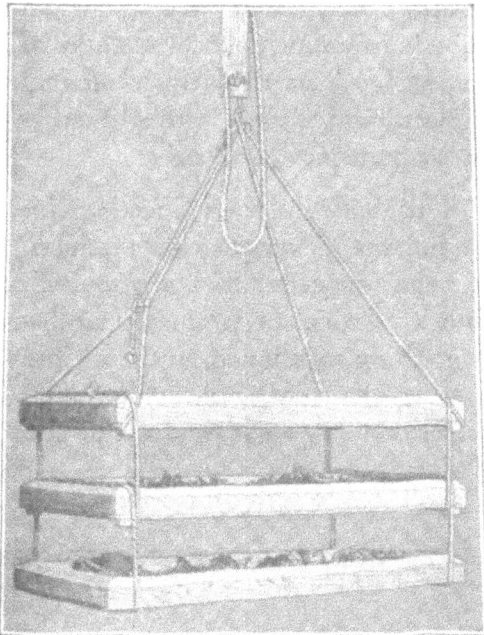

PURPOSE OF THIS BULLETIN

THERE are many ways of saving summer products for winter use. Of these, the drying of fruits and vegetables is but one.

In issuing this bulletin, it is not intended to suggest the drying of products in preference to canning, pickling, preserving or storing, but simply to present drying as an economical and practical method of saving food, particularly when any of the other methods cannot be easily or economically adopted.

This bulletin gives directions for drying many varieties of products. It is not expected that anyone will find it practical or profitable to dry everything mentioned, but from the long list given each housewife will be enabled to dry such products as she desires.

HOME DRYING

of

FRUITS and VEGETABLES

Compiled and Edited
By Edgar W. Cooley
of the
Agricultural Extension Department

NOTE—All or any portion of this booklet may be
reproduced by giving proper credit to the publishers

Published and Copyrighted 1918 by
INTERNATIONAL HARVESTER COMPANY
(Incorporated)
AGRICULTURAL EXTENSION DEPARTMENT
P. G. Holden, Director
HARVESTER BUILDING, CHICAGO

A E 28A–7-16-19

Advantages of Home Drying

HOME drying can be done by anyone, any-where, without extra expense for fuel or equipment and with very little extra work.

1. The process is simple—any boy or girl can make the equipment and do the drying.

2. No extra fuel is required—the drying can be done with the same heat used in carrying on the ordinary household duties.

3. Dried fruits and vegetables make whole-some food.

4. Drying preserves the flavor of the product.

5. Drying saves the product, saves storage space, saves transportation. The product is stored in ordinary paper sacks.

6. To dry food products is to save them for food. It is therefore economical — good business.

HOME DRYING

Food always has been and always will be an important problem; at this time, because of the great decrease in world production, as a result of the war, it is more than ever a matter which demands the serious consideration of every person.

Not only is it a question of maximum production, it is a matter of preservation of products and the elimination of waste. Wastefulness, bad at any time, today is a crime against society.

Saving of food should be practiced in every home. Every year thousands of home gardens are grown. Vegetables thus produced, and all other products which would otherwise be wasted, should be preserved by drying, canning or storing in pits.

The situation emphasizes the fact that waste is bad management; saving is profitable. Those who would provide themselves with plenty of fruits and vegetables next winter at a minimum cost, must eliminate waste now and can or dry the surplus.

First bottle contains fresh peas; second bottle same peas dried; third bottle same peas after being restored

We Can Save Everything

There isn't anything grown in the garden or orchard that we cannot save in some way. It can be pickled, or dried, or canned. It can be buried in the ground or in sand or sawdust in the cellar, or it can be simply put in the cellar.

The over-abundance produced in the summer should be the normal supply of the winter, and the individual family should conduct drying on a liberal scale.

Winter buying of vegetables and fruits is exceedingly costly, as you pay for transportation, cold storage and commission mer-

The process of home drying described in this bulletin was developed by H. S. Mobley of the Agricultural Extension Department, International Harvester Company, (Inc.), and has been used and demonstrated by him during the past 15 years.

chants' charges and profits. Summer is the time of lowest prices and summer is, therefore, the time to buy for winter use.

Home drying or canning of vegetables enables us to save what otherwise would be wasted in the home garden. For example: When we gather peas or beans, we should pick all that are in condition, and if the surplus is too small to market, we should dry or can it.

Drying can be done at little or no extra cost for fuel, if we utilize the heat used in other home work.

Any one can dry fruits and vegetables.

It is our duty to dry, can or preserve in some manner everything that would be wasted.

Little Storage Space Required

Drying was generally done by our ancestors but has been little practiced in recent years. It is important and economical in every farm, town, village or city home. To the city dwellers it has the special advantage that a great deal can be stored in a small space and at little or no expense. One hundred pounds of vegetable food can be reduced to 10 pounds by drying.

Any Food Product Can Be Dried

Many things are better canned than dried; others are better dried than canned. But if we cannot get cans or jars enough for canning there is scarcely any food product that cannot be dried in the home with no other equipment than what every family possesses or can easily make.

Drying saves the product, saves storage space, saves transportation. Dried products can be shipped anywhere—to hot countries or to cold countries. They may be kept anywhere, so long as they are in air-tight containers and are out of the reach of rats or mice, and will keep as long as the air does not reach them.

Anyone who will carefully follow the simple directions can successfully dry and save any product. When desired for food

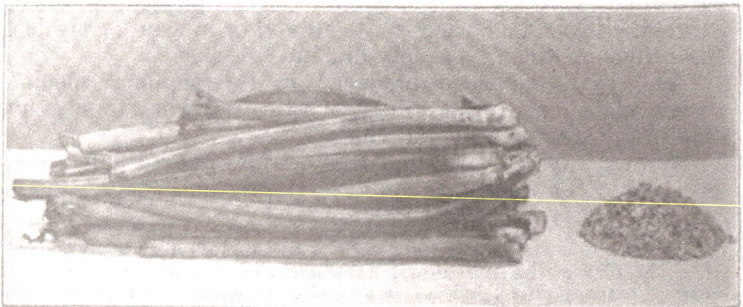

Three pounds of rhubarb are reduced to a few ounces by drying

all products can be partly restored, many of them to nearly their original condition.

The housewife who takes vegetables fresh from the garden and follows directions, being careful not to use too much heat in drying, will find, when she restores the dried product, that she has preserved all the taste and nutrition originally contained in the green vegetables.

Some of the points in drying cannot be explained—they can be learned only by doing the work yourself. But do not be confused or discouraged by this for the process is very simple and it will be a matter of surprise to you how easily it can be done and the different conditions overcome.

The products that can be dried are almost innumerable. The list embraces everything from the garden and orchard, including tomatoes, watermelons, and cantaloupes. It includes such highly perishable products as eggs, cream, cottage cheese and lean meats.

Expensive Equipment Not Necessary

The equipment for drying is as simple as the method. It consists of three frames such as any boy can make, any kind of a cook stove; a pot or pan or a tin bucket; a wire basket, or a flour sack, or even a piece of cheese cloth that can be fashioned into the shape of a bag by bringing the four corners together; a few pie pans, some dinner plates and an earthenware dish or jar.

Each of the frames should be about 27 inches long, 14 inches wide and 1½ inches deep. The sides and ends can be made of wood, and the bottom should be of galvanized window screen wire, fastened with double pointed tacks. The rack and harness shown in the cut were made by an eight-year old boy and what a young boy can do, anyone can do.

The heat from the stove goes up readily from the center and if the racks are too long, the product in the ends will not dry as

Two pounds of cottage cheese before drying; after drying weighs six ounces

rapidly as the rest. Just make the rack to fit the stove.

We must keep the racks clean. If they are any wider than 14 inches they cannot be placed in an ordinary dishpan. When you are through using the racks hang them on a nail in the pantry.

How to Make a Drying Rack

Get two pieces of small sized rope, or window weight cord, each six feet long. Tie the ends of each piece together, making two loops, each exactly 30 inches long. (Figure 2.) Place one loop around one end of the frame; the other loop around the other end (Figure 4). Bring the upper ends of the loops together and fasten them with a third loop, or doubled rope, sufficiently long to reach from a few feet above the stove to a firm hook in the ceiling (Figure 3). Near each end of a block of wood 8 inches long and $1\frac{1}{2}$ inches wide, bore a hole large enough to let the doubled rope pass through easily. Pass the end of the upright rope through one hole and shove the block down to the junction of the two loops (Figure 3). Tie a knot in the upright rope to keep the lower end of the block from slipping up; then pass the double rope through the upper hole in the block (Figure 1). Place the upper end of the upright loop over the hook in the ceiling.

The purpose of the block of wood is to make it easy to adjust the height of the frame. To raise the frame, pull the rope through the upper hole in the block until the desired height is reached, then fasten the frame in place by looping the "slack" of the rope around the upper end of the block as shown in Figure 5.

Place two loops of rope, each about 20 inches long, around the suspended frame, one loop at each end, and let them hang down. In the lower ends of these loops, place the second frame and suspend the third frame from the second in the same manner as shown in cut on page 10.

Three Frames Save Time and Fuel

By using three frames the housewife will be able to save both time and fuel in drying a quantity of vegetables. For instance if she is drying peas, she can prepare enough to fill one frame and let them be drying while shelling, blanching and cold dipping another batch. When the second frame is spread with peas, those placed in the first frame will have been drying for about 25 minutes. This frame can then be raised until the second frame, suspended from it, is the same distance above the stove the first frame had been. When the third frame of peas are ready to dry, the second will have been drying about 25 minutes and the first about 50 minutes, but more slowly. Frames 1 and 2 are again raised until No. 3 is suspended the right distance from the stove.

When the peas in the third frame are sufficiently dried, those in the second frame, which have been drying for 25 minutes longer

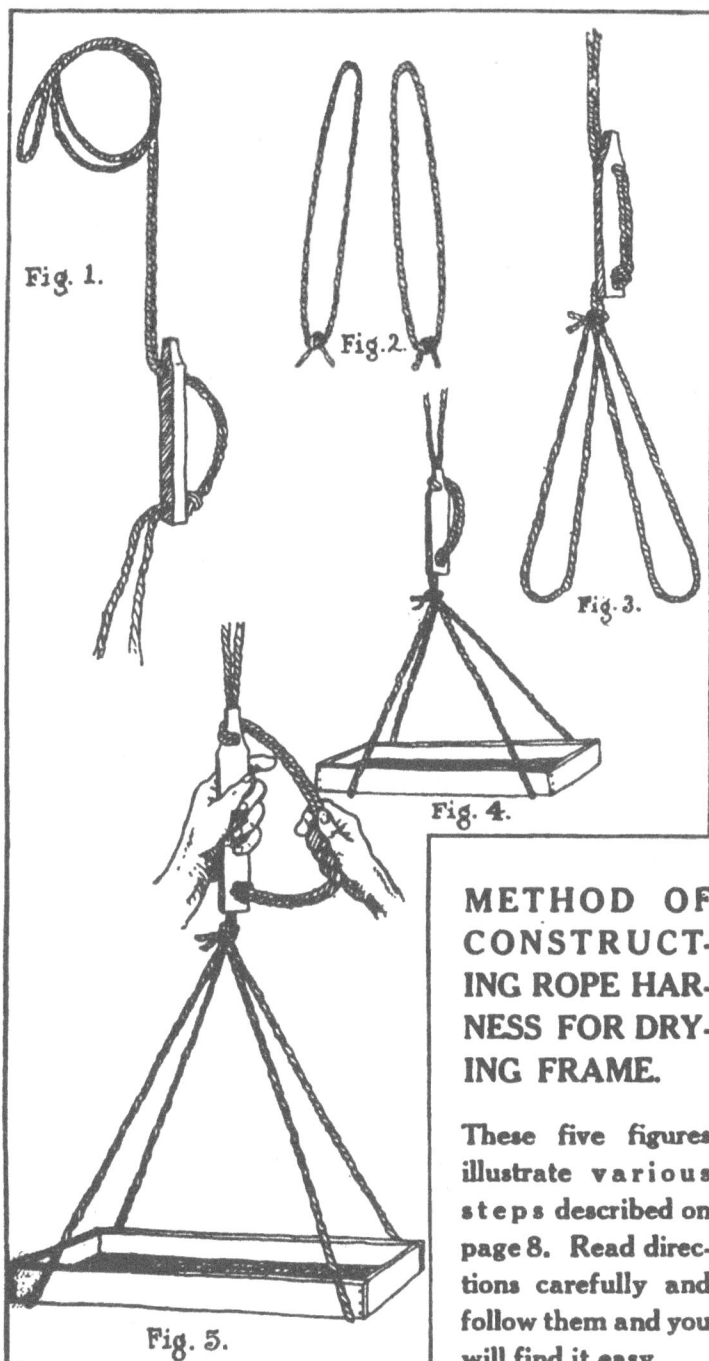

Fig. 1.

Fig. 2.

Fig. 3.

Fig. 4.

Fig. 5.

METHOD OF CONSTRUCT-ING ROPE HAR-NESS FOR DRY-ING FRAME.

These five figures illustrate various steps described on page 8. Read direc-tions carefully and follow them and you will find it easy.

but the most of the time at a lower temperature, and those in the first frame, which have been drying 50 minutes longer but at a still lower temperature, will have been dried to about the same extent. All the frames can then be removed and emptied and the process begun over again. Never start the three racks to drying at the same time.

In drying products no iron bound rule can be followed, as conditions depend upon the weather, the maturity of the product, the amount of moisture in the product, whether the products where gathered in the morning or in the afternoon, whether they are fresh or wilted, and whether they are large or small in size.

Circulation of Air Important

You must remember you can scarcely find two batches of vegetables and fruit exactly alike and this fact varies the time necessary to dry them. Much depends, also, upon the regularity with which the heat is applied. The main thing is to be sure there is enough air circulating or the vegetable will never get dry. The air must circulate freely.

In using coal or wood for fuel, never take the lids off the stove. In using gas or coal oil, if there are lids on the stove remove them and always keep the flame as low as possible. You will be surprised at how low the flame can be used.

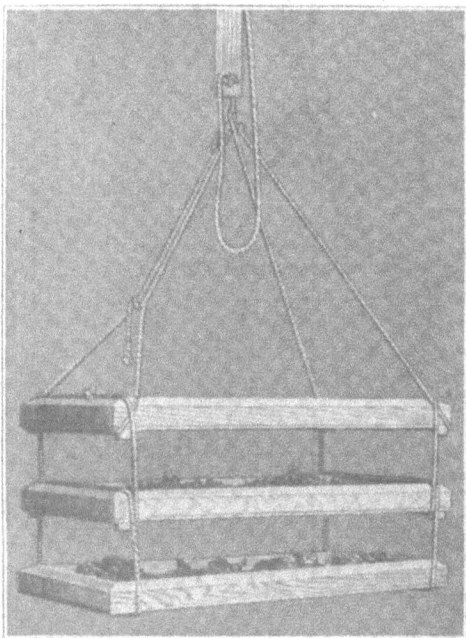

Method of suspending three drying frames with rope harness

In drying any kind of products the same general process is followed, although the details may vary. But to maintain the original taste of the fruit or vegetable, drying should never be done at a higher temperature than 120 degrees Fahrenheit. All experiments have shown that if a greater heat is used, there is danger of burning the product and the delicate taste is sure to destroyed.

Too Much Heat Injures Flavor

Here is a simple test for determining whether the rack is too hot: Place the palm of your hand against the under side of the bottom of the rack. If you are inclined to jerk your hand away, the rack is too hot and it should be raised to a greater distance from the stove. Don't be afraid of using too little heat but always be afraid of using too much heat.

Drying in the sun is all right but it takes too long. A product that will dry over a stove in three hours will require three days to dry in the sun.

There is no rule by which to ascertain when products are thoroughly dried. Some think products when sufficiently dried will rattle when you lift them up, but this is not a safe test in all cases. The best way to judge is to find out by experience— by drying something until you think it is dry enough and then examining it every few days to see if it is molding. The cracker test is a fairly reliable one. Put a cracker in the bag for a few days when you put the product away. If the product is not dry enough, the cracker will be moist. But remember all tests fail. You must learn by trying and you will find it easy. There is little danger of your drying anything too much if you do not use too much heat, follow directions and use judgment.

If you are called away from the kitchen for a considerable length of time and the article being dried is not yet dry, turn out the burner, put a cloth over the frame and leave it. When you return go on with the drying. Do the same way if you leave it overnight. The product should be covered with a cloth to keep moths and flies off.

During the drying process all products will appear to stick to the rack, and the person doing the drying will be inclined to pull the product loose with the fingers. Some people fail for that reason. The product should be let alone until thoroughly dry, when it will come loose very readily.

Brief directions for drying a number of different products follow and from these the housewife can determine how to dry any fruit or vegetable by following the method employed in drying products of like nature.

Vegetables, like corn, beans, peas, cabbage, potatoes, carrots, and parsnips should be blanched.

They are blanched by being placed in a wire basket, a flour sack or in a piece of cheese-cloth or towel, the ends of which have been twisted together to form a sack, and then placed in boiling water for about eight minutes. They must then be removed and cold-dipped by plunging them at once into cold water and letting them remain there one minute.

Sweet corn should be blanched on the cob, then the kernels cut off and spread to the depth of about one-quarter of an inch upon the bottom of the drying frame.

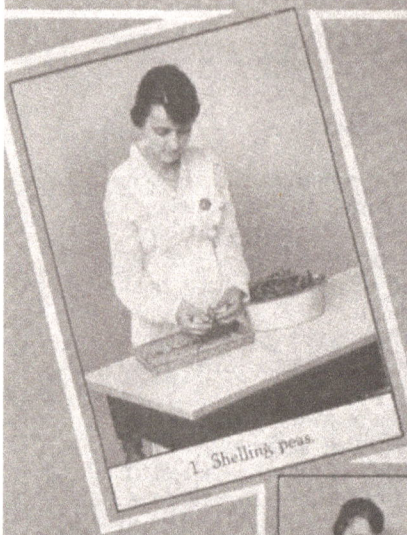

1. Shelling peas.

2. Placing peas in boiling water to blanch.

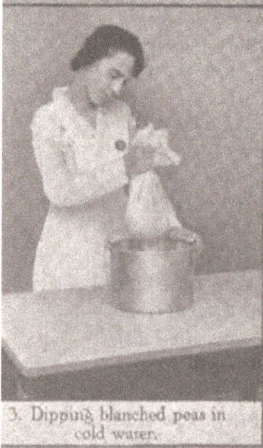

3. Dipping blanched peas in cold water.

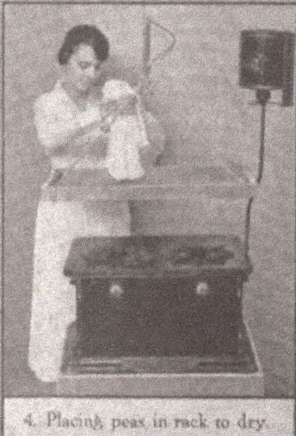

4. Placing peas in rack to dry.

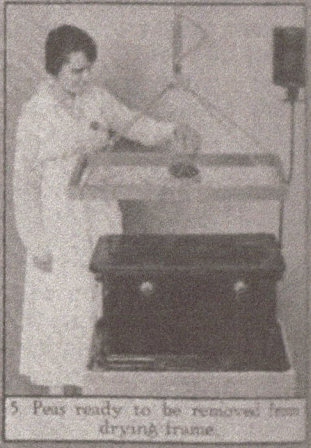

5. Peas ready to be removed from drying frame.

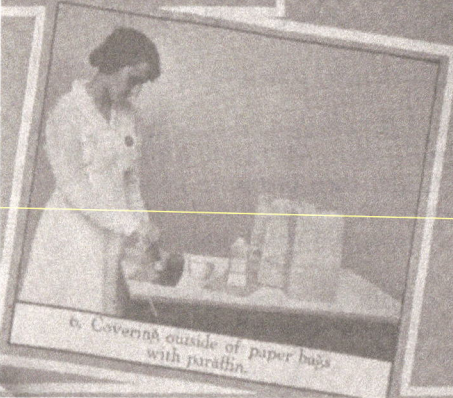

6. Covering outside of paper bags with paraffin.

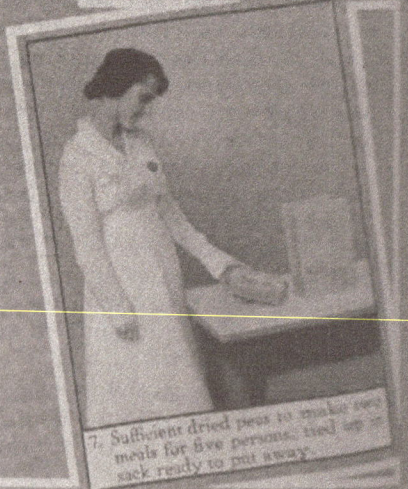

7. Sufficient dried peas to make two meals for five persons, tied up sack ready to put away.

Carrots, parsnips and **potatoes** should be scraped and sliced before blanching, then carefully drained of moisture and placed in the drying frames. Only very tender carrots should be dried.

Squash and **pumpkins**—Cut into inch slices, peel off rind, chop into pieces ¼ inch thick. Spread in rack and dry.

Beans—In preparing and selecting product follow method employed in Cold Pack Canning. Never dry tough beans or beans with very much string on the pod. Break off tip ends, blanch, and dry whole, pods and all. If some pods are tough, do not throw them away. Hull them and dry the beans.

Peas—Hull before blanching. Spread on rack to dry.

In drying peas you will have some little peas and some big ones. The little ones will shrivel up; the big ones won't. Sort the peas. Put the big ones in one tray and the little ones in another tray. Naturally the smaller the thing, the quicker it will dry. Five quarts of peas in the pod will be sufficient, w h e n they are hulled, to fill a frame.

Cabbage can be dried but it is better to preserve it in the form of kraut.

The process to be observed in drying other vegetables and fruits follows:

Beets — Do not blanch as blanching causes them to bleed and they lose some

Pint of cream in original state and the quantity in dried form

of their nutritive value. They should be peeled, washed, sliced, and laid in the frame to dry.

Rhubarb—Do not blanch. Wash, drain, slice in small pieces and dry. Rhubarb does not need to be peeled if it is tender. In vegetables, apples, and peaches, the mineral salt and most of the nutrition are next to the peeling and if we peel them we destroy some of these qualities.

Greens—Do not blanch. Wash, drain off moisture and dry whole. The only exception to this is that Swiss Chard or any other greens having a thick stem, should be cut up into half-inch pieces.

Asparagus—Cut off all that portion that would be tough when cooked. Cut the remainder into ½ or ¾-inch lengths and dry without blanching.

Tomatoes—Select firm and ripe, not watery fruit. Wash, slice, lay in rack and dry.

Plums—Wash, remove pit, cut into quarters, spread in rack.

Cherries—Wash, remove pit, dry whole.

Strawberries—Spread in racks and allow to remain until no moisture comes from the berry when it is mashed between the fingers. Large berries may be cut in two. Dry all other berries in same manner.

Wild fruits—Use same process as in drying cultivated fruits. Persimmons, figs and the old "Possum Apple" can also be dried.

Miscellaneous Products

The process of drying cream, cottage cheese, eggs and meats is equally simple.

Cream—Cover the bottom of a pie pan to the depth of about a quarter of an inch with the cream; set in rack and dry about eight hours or until you can see the oily cream is free of all water.

Cottage cheese—Cover the bottom of the rack with cheese cloth; spread the cottage cheese on the cloth to the depth of about a quarter of an inch; dry for about four hours, or until the cheese becomes yellowish and grainy.

Eggs—Break the eggs into a crock or dish and beat until the whites and yolks are thoroughly mixed; pour into pie-pans to the depth of a quarter of an inch. Set pans in rack and let dry until egg forms a thick paste. Run paste through a meat chopper and grind it to a putty-like powder. Put the powder back into the pie-tins and dry for about an hour.

Meat—Any kind of lean meat—not fat meat—can be dried. Cut up the meat and grind it in a meat chopper; spread on a piece of cheese-cloth and place in rack to dry.

All products, if properly dried, will keep indefinitely if placed in air-tight containers, which need not be sterilized. Glass jars or bottles, should be thoroughly washed.

Vegetables and fruit, except watermelon, can be put away in paper bags that have been made absolutely air-tight by an application of paraffin.

Cut up two ounces of paraffin and dissolve it overnight in eight ounces of gasoline. With a small paint brush cover the bags all over on the outside, with the paraffin. Let the bags dry for two days in the open air before using.

Place in Sacks

Put in each sack enough of the dried product for two meals for your family; tie up sack so it will be air-tight as shown in cut. Hang up anywhere or put on shelf out of the way of rats and mice.

The inexperienced housewife putting dried products away in bags, should examine them every three or four days for about two weeks, or until she knows they are keeping. If any mold shows, the product should be immediately spread in the rack and dried some more.

Eggs, cream, meat and watermelon should be put away in glass, never in paper. Use four or five-ounce wide-mouth bottles and keep tightly corked. Put cottage cheese away in paper bag.

It is not safe to believe that when products are put away in glass, moisture will show on the glass if they are not thoroughly dry. The moisture may be caused by other things.

To restore dried products it is well to remember that the longer the article has been dried, the longer it should be soaked. A good method is to spread product on a level pan or plate and barely cover with water. Once soaked until they have been restored to about their original condition, dried vegetables and fruit can be cooked in about the same manner as though they were fresh. Try cooking in the same water in which they were soaked, with a little additional water added. Also try cooking in fresh water and use the method which the better suits your taste.

Most dried vegetables should be cooked rather slowly. Try cooking both ways—slowly and rapidly—and decide the way you like better.

Follow no set rule. Rather take an interest in your work, use your judgment and acquire skill in finding a method of your own.

To Restore Products

The best process of restoring various products is as follows, it being understood that the relative amounts of water and product given here are only approximate and will vary according to

Five quarts of strawberries before and after drying

conditions. Observation and judgment will easily determine the amount of water needed in each case.

Snap-beans—Soak from eight to 12 hours in 10 pints of water to one pint of dried product.

Beets—Soak two hours in two pints of water to one pint of product.

Corn—Soak from two to four hours in two pints of water to one pint of product. If soaked longer than four hours keep product very cool as there is danger of its souring.

Irish and sweet potatoes—Soak from six to eight hours in eight pints of water to one pint of product.

Rhubarb—Soak from six to eight hours in 12 pints of water to one pint of product.

Spinach and other greens—Cook slowly without soaking, or soak two to six hours. Try both methods and follow the one that suits you better.

Okra—Soak until soft.

Onions—Cook slowly without soaking.

Carrots—Cook slowly. No soaking necessary.

Parsnips—Soak two to four hours, using two parts of water to one of product.

Squash or **pumpkin**—Soak eight to 12 hours in 10 parts of water to one of product.

Turnips—Use eight parts of water to one part of product. Bring slowly to boiling point, boil about 20 minutes.

Cherries—Soak six to eight hours in four parts of water to one part of product.

Strawberries, blackberries, raspberries, etc.—Soak four to five hours in six parts of water to one part of dried product.

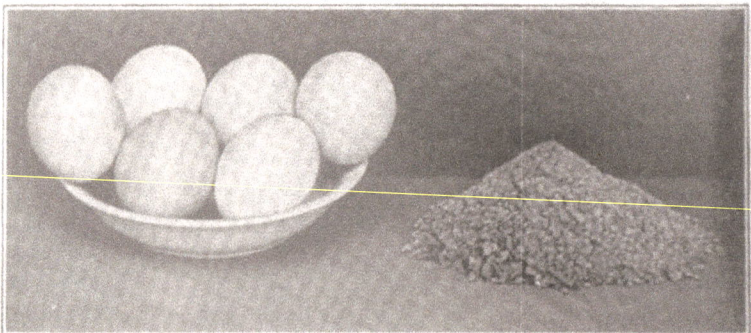

Half dozen eggs before and after drying

Eggs—Put in earthen vessel, cover with water, but not enough to make product float; let stand over night. Can be used in any manner in which eggs are used, except for poaching, boiling, or in any cooking where the white and yolk are used separately.

Cottage cheese—Cover flat pan ¼ inch deep with product and barely cover with water. Let it stand two hours. Do not use milk to restore it as it is only the water that has been evaporated.

Cream—Can be used in its dried form for cooking and seasoning.

THE FOLLOWING IS FOR HISTORICAL PURPOSES ONLY
AND DOES NOT REFLECT CURRENT
SCIENTIFIC KNOWLEDGE, POLICIES,
PRACTICES, METHODS OR USES.

I H C

AGRICULTURAL LECTURE CHARTS

REFERENCE BOOK
FOR THE
LECTURE

HOME CANNING

BY THE

COLD PACK METHOD

PUBLISHED BY THE

INTERNATIONAL HARVESTER COMPANY OF NEW JERSEY, Inc.

AGRICULTURAL EXTENSION DEPARTMENT

HARVESTER BUILDING, CHICAGO

PRESERVE YOUR HOME   117

HOME CANNING

BY THE

COLD PACK

METHOD

Cold Pack Canning Means: Packing the Food in the Jar Uncooked, and Then Cooking It in the Closed Jar

---

Prepared by Grace Marian Smith

PUBLISHED AND COPYRIGHTED 1917, BY
INTERNATIONAL HARVESTER COMPANY
OF NEW JERSEY (INCORPORATED)
AGRICULTURAL EXTENSION DEPARTMENT
P. G. HOLDEN, DIRECTOR
HARVESTER BLDG., CHICAGO

MUCH food is in the tillage of the soil: But there is that which is destroyed for the want of judgment.

Prov. 13-23.

INTRODUCTORY

FOOD always has and always will be an important problem; at this time because of the war it is more than ever a matter which demands the serious consideration of every citizen. It is not only a matter of increased production, but also of economy in consumption, preservation of products and the elimination of waste. Wastefulness, bad at any time, today is a crime against society.

Saving of food should be practiced in every home. This year thousands of home gardens have been grown. Vegetables thus produced and all other products which would be wasted, should be preserved by drying, canning, or storing in pits.

Canning by the Cold Pack method is simple and inexpensive. You can successfully can vegetables and fruits of all kinds. There is no excuse for wasting products of the garden, field and orchard. Can them. Our people will need them at home—our soldiers will need them at the front.

In this publication we have endeavored to describe the steps in Cold Pack Canning in a practical way so that it can be done by anyone in the home or by canning clubs.

In the preparation of this book many have assisted with suggestions. Special acknowledgment is due O. H. Benson, U. S. Department of Agriculture.

TABLE OF CONTENTS

Page

Suggestions for Conducting Canning Meetings and Demonstrations.................................... 5
Illustrations—Demonstrator's Outfit...................... 6
Chart I—Don't Waste It; Can It...................... 8
Chart II—Why Can It.............................. 9
Chart III—Anyone Can Successfully Can Any Product by the Cold Pack Method 11
Chart IV—Use What You Have...................... 14
Testing Jars and Rubbers...................... 15
Use the Cooker You Have.......................... 17
Types of Factory Made Outfits...................... 19
Other Things Needed........................... 22
Chart V—Steps in Cold Pack Canning 23
Illustrations—Steps in Canning 24
Canning Tomatoes............................... 26
Chart VI—Finishing the Work...................... 30
Chart VII—It's Good Business...................... 33
Chart VIII—We Grow It, Why Not Can It?.............. 34
Chart IX—Club Work Gives 4-H Training.............. 35
Chart X—Why Have I Been Talking to You About Home Canning by the Cold Pack Method?............ 38
Opportunity for Boys and Girls 40
The Mother's Responsibility....................... 41
Canning in Tin.............................. 43
The Hand Sealer............................. 44
Sterilizing Products in Tin...................... 45
Labels 46
Buying Food to Can........................... 46
Solder Sealed Tin Cans.......................... 47
Canning Reminders.......................... 51
Canning Fruit Juices and Meats.................... 55
Jellies and Preserves............................ 56
History of Home Canning Clubs.................... 58
Save—Don't Waste............................. 59
Time Table................................. 60
Canning Literature 61
The Visual Method of Instruction.................... 62
Educational Publications.......................... 63

4

SUGGESTIONS FOR CONDUCTING CANNING MEETINGS AND DEMONSTRATIONS

Arrange for a short lecture and if possible a demonstration of Cold Pack Canning at some school house, church, or other suitable place. If it is not possible to give the demonstration at the meeting, instruct in the Cold Pack method and endeavor to arouse interest in home canning. An organization should be effected and a canning demonstration arranged for a later date.

Publicity—Advertise the meeting thoroughly in the local papers, at schools, churches, granges, and other meetings by hand bills and posters.

Products to Can—Ask local people to bring fruits and vegetables, but be sure to provide yourself with sufficient products. This will insure a supply.

Do Not Try to Can Too Much—Demonstrate *method*, not quantity. Three kinds of products for canning are enough; never use more than four. A small number avoids difficulty in arrangements and concentrates attention. Can one kind of fruit, one kind of vegetable, and tomatoes. If you can a fourth product, select corn, beans, greens, sweet potatoes, pumpkin, or some other vegetable not commonly canned at home. Do not attempt to can any product when the time required is greater than that given for the meeting.

Demonstrator Must Have Assistants—It is always advisable in giving demonstrations to have at least one helper, who should be sufficiently familiar with Cold Pack Canning to direct volunteer helpers, and keep things moving. This enables the lecturer to devote his time to making clear the steps in canning and show the importance and advantages of the canning work. The demonstrator or his assistant should see that the PACKING is properly done, rubbers and covers properly fitted, and jars placed in the cooker. Avoid mistakes. Mark the time on the blackboard, if there is one in the room.

Canning Outfit—Glass jars and a home-made hot water outfit should be used for the first demonstration. This convinces the audience that an elaborate equipment is not necessary. With a hot water outfit each jar can be set in as packed, but if a steam pressure outfit is used, the cooker cannot be opened after cooking begins. It is an advantage to have a steam cooker for corn, sweet potatoes, and other vegetables that require long cooking. This will make it possible to finish cooking such products and exhibit them at the close of the meeting. Do not fail to have a fire extinguisher. Any good make will do.

Get Action—Arrange with a local worker to make a motion at the close of the lecture, providing for the appointment of a committee to report on the organization of a canning club and to take steps to secure a county canning club leader. A canning demonstration or lecture should never be considered an entertainment. It should give information, arouse interest, be directed toward some definite work for the year.

Canning School—A Canning School lasting an entire day or several days may be planned. Arrangements for such a school should be made several days in advance so that heat, water, seats, utensils, jars, seasonings, and products are sure to be provided.

A DEMONSTRATOR'S OUTFIT

The demonstrator should carry the articles illustrated below for community and county work. The community in which the meeting is held should furnish two gasoline stoves with two large burners each, matches, gasoline, a long table, chairs, a pail, dishpan, jars, salt, sugar, and product to can.

Two 2-burner stoves are more convenient than one 4-burner.

The pans and pan lifter pack more easily than a stewpan with a handle. If the jar holders cannot be secured, a lath rack as described on page 17 may be substituted. The rack may also be used when blanching in steam.

It is wise to have on hand some small rubber bands, some heavy cord, tacks, nails, hammer, wrench, heavy paper or oilcloth for covering table, and if canning in tin is to be demonstrated, a supply of cans, solder, flux, sal ammoniac, capper, tipper, hand sealer, and sanitary seal cans. **Do Not Fail to Have a Fire Extinguisher.**

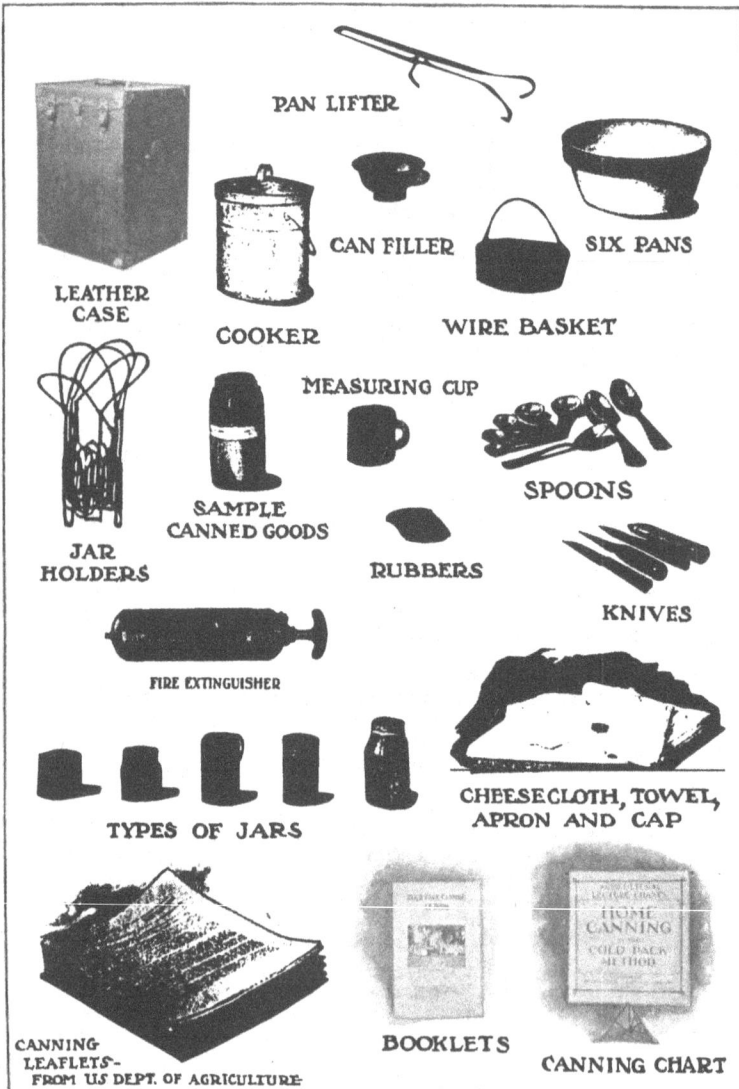

PAN LIFTER

CAN FILLER

SIX PANS

LEATHER CASE

COOKER

WIRE BASKET

MEASURING CUP

SPOONS

SAMPLE CANNED GOODS

JAR HOLDERS

RUBBERS

KNIVES

FIRE EXTINGUISHER

TYPES OF JARS

CHEESECLOTH, TOWEL, APRON AND CAP

CANNING LEAFLETS— FROM US DEPT. OF AGRICULTURE

BOOKLETS

CANNING CHART

HOME CANNING
BY THE COLD PACK METHOD

Every year in the United States we waste vegetables and fruits sufficient to feed thousands of families. Products of the garden, field and orchard rot upon the ground where they are produced, while there is abject want and suffering in many sections throughout the land. In thickly populated centers hunger is a problem which has for years demanded the attention of charity. In our cities last year there were bread riots. At that time we were not engaged in war. What will this fall and winter bring forth, now that we are sending an army to the front? In America the crime of waste must be stopped. To save food is a duty, an obligation of citizenship. Why not answer the nation's appeal to do

COVER CHART

our part? We will need all that the land can produce, and we can save it by canning, drying, burying in pits, storing in cellars, preserving the fruits in the form of jellies and jams. A simple and inexpensive process of preserving the surplus food stuffs is by the Cold Pack Method of Canning. Fruit and vegetables of all kinds can be preserved in this way.

On the pages that follow are the steps in Cold Pack Canning. Follow the directions closely. You will have no trouble. The boys and girls of the Home Canning Clubs of the country are canning thousands of jars of food products by this method.

DON'T WASTE IT; CAN IT

Three-fifths of a ton of tomatoes (1200 pounds) is an average crop from $\frac{1}{10}$ acre, a piece of land 66 feet square.

DON'T WASTE IT CAN IT

A CAN OF FRUIT
A CAN OF GREENS
A CAN OF VEGETABLES
FOR EVERY DAY IN THE YEAR

CHART I

Tomatoes sell fresh at $8 to $10 per ton. Suppose the crop from $\frac{1}{10}$ acre is worth $6 wholesale. Three-fifths of a ton of tomatoes canned will make forty dozen quarts, and at 15 cents a can, the price we pay for them at retail, they are worth $72. Is it not worth while?

"A Can of Fruit, a Can of Vegetables, and a Can of Greens, for Every Family, for Every Day in the Year" when the garden is not producing—this is the slogan of the Home Canning Clubs of the United States.

Canning Demonstration Before the Waveland, Ind , Woman's Auxiliary

8

WHY CAN IT?

Gives Greater Variety — Is Wholesome — Saves Doctor Bills—We Like It Better. Canned foods retain the natural juices and flavors, and in addition to being nutritious and healthful, are tasty. We like them.

A large percentage of the medicines sold are patent laxatives. We could do without most of the patent laxatives if we ate more fresh and canned fruits and succulent vegetables.

Why should we limit our diet to meat, bread, and potatoes three times a day, when tons of fruits and vegetables go to waste?

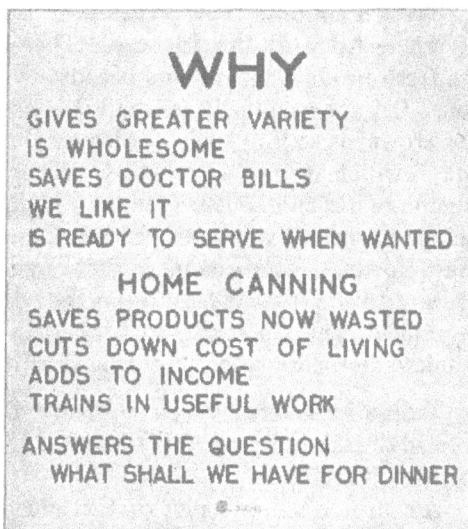

WHY

GIVES GREATER VARIETY
IS WHOLESOME
SAVES DOCTOR BILLS
WE LIKE IT
IS READY TO SERVE WHEN WANTED

HOME CANNING

SAVES PRODUCTS NOW WASTED
CUTS DOWN COST OF LIVING
ADDS TO INCOME
TRAINS IN USEFUL WORK

ANSWERS THE QUESTION
WHAT SHALL WE HAVE FOR DINNER

CHART II

Our efficiency depends on what we eat. An unbalanced diet means slow, stupid, headachy, ill-tempered people. If we canned more fruits and vegetables, we would eat more of these and less of the heavy foods.

Is Ready to Serve When Wanted — In an emergency—a request at eleven o'clock to have dinner an hour early; unexpected company arriving just as dinner is served; the housewife coming home late after a day spent shopping, or calling, or at church — think of the comfort of knowing that there are on the shelves: home-canned soups, meats, vegetables, greens, fruits, and fruit juices.

If she keeps her shelves well stocked, any housewife can prepare a good dinner any time, in the few minutes required to open the jars and heat the products.

9

HOME CANNING SAVES FOOD

By Home Canning we mean the canning commonly done by the housewife, and also canning done by the Boys' and Girls' Clubs, in backyards and on club plots.

Saves Products Now Wasted — Cuts Down the Cost of Living —Adds to the Income. There may be no demand for the fresh products near home because everyone grows some garden. Or, the surplus is so small it is not marketed; or, even if the grower is selling some garden produce, there are the "seconds" which do not sell readily. Then there are the products which are difficult to can — those which would not keep when canned in the old way. Cold Pack Canning at Home will save these foods. Home canned fruits and vegetables reduce the grocery bill. It costs less to can them than to buy them, fresh or canned, and they also cost less than the higher priced, less wholesome foods which might be substituted for them.

Trains in Useful Work — Every boy and girl should be trained to make a living. We learn to do by doing, not by reading how it is done.

A combination of a plot of Ground, a Club Member, and a Canning Outfit has great possibilities.

Home Canning Answers the Question, "What Shall We Have for Dinner?" and answers it in a way that gives a varied menu for every day in the week, and helps make Sunday *a real day of rest,* for Mother as well as for Father and the Boys.

Group of Club Leaders Training for Field Work. Canning Kitchen,
U. S. Department of Agriculture

10

ANYONE CAN SUCCESSFULLY CAN ANY PRODUCT BY THE COLD PACK METHOD

Cold Pack Canning means: Packing the Uncooked Food in the Jar and then Cooking it in the Closed Jar.

By this method it is possible for anyone to can at home, in one process, any food product and know that it will keep.

Scalding is a familiar term; in canning it is understood always to mean immersing in boiling water or steaming.

Blanching is more commonly known as parboiling. It means that the product is left in the boiling water, or the steamer, for a longer period than is indicated by scalding. The time varies for different products. (See pg. 60)

Cold-dip means plunging at once into cold water and out again.

ANYONE CAN CAN ANY PRODUCT
BY THE COLD PACK METHOD

HOW IT IS DONE
SCALD OR BLANCH
AND COLD DIP ALL VEGETABLES
PACK THE PRODUCT UNCOOKED
CLOSE THE JAR
COOK IT IN THE CLOSED JAR

WHY COOK IN JAR
THE PRODUCT IS BETTER
IT STERILIZES COMPLETELY
PREVENTS ANY BACTERIA GETTING IN
SAVES WORK AND TIME
TAKES THE DRUDGERY OUT OF CANNING

IT IS THE ONLY SURE WAY

CHART III

Do not neglect blanching all vegetables — most fruits should not be blanched. Blanching eliminates objectionable acids and acrid flavors. It also shrinks the product, which allows more to be placed in the jar.

All vegetables must be blanched and cold dipped. Many of the fruits do not need blanching, but those which are scalded or blanched *must be cold dipped at once.*

Pack the Product Uncooked —Of course, such products as are blanched or scalded are heated a trifle; but many of the fruits are packed fresh without blanching and in every case the real cooking is done in the jar.

11

Close the Jar—If we are canning in glass we do not seal the jar but close it lightly. Heating the contents causes steam to form and if no outlet is provided, the pressure of the expanding steam might be sufficient to break the glass.

If we are canning in tin, we seal the can tight. The tin will give enough to allow for the expansion, and as the contents cool, the can will return to its original shape.

Cook It in the Closed Jar—Cooking, sterilizing, or processing as it is called in commercial use, means heating to the point necessary to keep. The time required for cooking varies with the kind of product and the kind of outfit. (See time-table, page 60.)

WHY COOK IN THE CLOSED JAR?

It Sterilizes Completely — Prevents Any Bacteria Getting In—If the product is put into the jar, the jar closed, and the product cooked in the closed jar, we are certain the organisms which were present are killed; and the sealed jar prevents any bacteria which may be in the air from getting in after the product has been cooked.

By the old open-kettle or hot-pack method, it is impossible to know that any given jar or product is perfectly sterilized. Even when the products, the jars, the rubbers, and the covers have been sterilized there is still danger of bacteria getting in *while the cooked product is being dipped from the kettle into the jar.*

The Product Is Better — It is better in color, in flavor, and in texture. It is not crushed, nor cooked until it is mushy; instead of a conglomerate mass, each berry or slice is distinct.

It Saves Work and Time—By this method it is only a trifle more work to can a half bushel than it is to can a quart.

"Old Man" Bacteria Can't Get In

Once the product is prepared and put into the jar, it is as easy to cook a dozen jars, if the cooker is large enough, as it is to cook one, and it requires no more time.

It eliminates entirely the hot, trying work of dipping from the kettle to the jar.

When canning with the open kettle, it is just as necessary to sterilize the jars carefully, to test rubbers to fit tops, and to seal perfectly, at the last minute, with the very last jar, at the end of a long, hot, tiresome day when one is finishing a large lot, as it is the first hour of the morning.

By the Cold Pack method, the work which needs care is all done in the beginning when the worker is fresh.

Then we do not have to watch the pack all the time it is cooking. There is no danger of "burning the kettle."

Housewife Showing the Racks She Used — a Shelf from the Ice Box, and a Home-Made Lath Rack.

When cooking fruits or vegetables in the jar one needs only to note the time when boiling begins (or, if using a steam outfit, when the required steam pressure is reached) and the worker may then go about other work, setting an alarm clock to ring when it is time to take the product off the fire.

It Takes the Drudgery Out of Canning—We no longer dread the canning season. Canning by this method is an inter-

Still Doing His "Durndest" But Can't Get In

esting, business-like proposition; not drudgery. It is pleasanter to pack fresh vegetables in a cool room, than to pack hot vegetables in a hot room.

To sum up: By the cold pack method,

Anyone can can any food product—fruit, vegetables, meats, fruit juices, greens, soups, fish, game, or fowl.

The work is easier, pleasanter, and more interesting than by the hot pack, or the three-day, intermittent method.

The product is better, and, finally,

It is the Only Sure Way.

USE WHAT YOU HAVE

Special Canning Outfit Not Necessary

We can do Cold Pack canning with any style of glass jar or tin can, except those which are sealed with wax. Likewise the wash boiler, or a galvanized pail, or a large kettle may be used for a cooker.

CHART IV

Glass Jars, Tops and Rubbers—Imperfectly sealed jars are probably responsible for more spoiled canned goods than any other one cause. Before beginning to can, fit the tops to the jars, and test the rubbers. Wash the jars, tops, and rubbers in hot soap-suds and rinse in boiling water. If the tops are old, boil them in water to which a little soda has been added. If they cannot be cleaned so as to be perfectly sanitary and also to look clean and neat, do not use them—get new ones.

Place the jars and tops in a kettle of warm water and allow it to come to a boil. Leave them in the boiling water until you are ready to fill them.

Rubbers should not be boiled to sterilize them, but cleaned by washing in hot water to which a little soda has been added. Too prolonged heating injures the rubbers, and as they have to stand long boiling on the jars it is unnecessary to subject them to the extra strain.

Use new rubbers. Rubbers bought new from the store are not always new; they may have been carried over from last year's stock.

Rubbers which are extra thick and wide are not necessarily good rubbers. They may lack elasticity, they may be unnecessarily wide, or so thick they do not permit the cap of the can to screw down tight.

Testing Rubber Rings

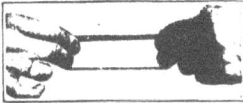

Turn and Stretch the Rubber

Buy as good rubbers as you can get, then test for elasticity by pulling one or more times to see if they return to shape and do not break. Turn and stretch the rubber so that all parts of it are subject to the strain.

Testing Jars and Covers

Screw Top Jars — Put the top on without the rubber; screw down as tight as possible. If the thumb nail can be inserted between the cover and the jar at any point of its circumference, either the cover or jar is defective. Sometimes the edge of the cover can be bent down to make the joint tight.

Testing the Cover

The Cover Should Fit So Tight That the Rubber Cannot Slip Back

Next, place the rubber on the jar, and screw the cover down with the thumb and little finger in the same way as when preparing the jar for cooking the product. Catch hold of the rubber and pull it out, and then let it fly back. If it slips into place under the cover, the cover is not a good fit and either the cover or the jar should be discarded. Third, run the thumb around the surface on

Testing for Rough Edges

Smoothing the Edge

which the rubber rests. If the edge of the jar or the cover is rough it will cut the rubber. Sometimes with a file or an old knife a rough edge may be rubbed smooth, using care not to turn the edge and spoil the seal.

Jars with Composition Attached to Cover—Set the cover on the jar and tap all the way around the edge to see if the cover sits level on the jar. If it rocks at any point, this indicates a defect in either the cover or the jar.

Testing a Vacuum Seal Jar

The composition attached to covers sometimes deteriorates with age, even if the cover has not been used. In buying covers with rubber attached, be sure they have not been carried over from last season. Old covers of this type should be thrown away.

Glass Top with Spring Clamp—Put the cover in place without the rubber, set the spring, and press the clamp down. If the thumb nail can be inserted between the cover and the jar, the spring is not tight enough. To remedy, disengage the ends of the top spring from the eyelets at the side. Holding a side of the bail in each hand, press down with the thumbs on each side of the top bar. This will cause it to fit closer to the cover and increase the pressure. Return the spring to the jar and test again. Sometimes the glass covers of these jars break because the spring fits too tight.

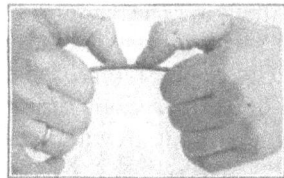

Bending the Bail to Tighten it

Jars with Wide Mouths—Jars with wide mouths, straight sides, and lacquered or glass tops are usually preferred. They clean and pack more easily, and will take large fruits and vegetables whole.

USE THE COOKER YOU HAVE

It is not necessary to have a special outfit to do Cold Pack Canning. A common wash boiler found in nearly every home makes a good cooker; it is deep enough, the sides are straight and it has a close-fitting cover; or a large galvanized pail with cover, such as used for garbage pails, will serve the purpose. For canning small quantities, a tin or galvanized water pail, lard can or coffee can are good.

Fig. 1 — Tray Made from a 10-cent Collapsible Vegetable Crate — Costs Nothing but a Little Time to Make It

No matter what kind of a cooker is used it must be at least three inches deeper than the tallest jar. This will give room for the rack upon which to set the jars and an extra inch and a quarter so it will not boil over. (See drawing on Chart IV, page 14.)

It is an advantage to have the cooker at least 13 to 14 inches deep. Jars must not set directly on the bottom of the cooker, or the contents will become overheated, and in overheating there is danger of the jars breaking. Also, when the jars become overheated part of the contents will escape under the cover and be lost. To prevent this danger, jar holders or trays are made of lath, wire, tin or perforated board. The tray should rest on the slats so that it will be raised an inch above the bottom of the cooker. This will allow water to circulate freely around the bottoms of the jars.

The tray shown in the illustration on this page (See Fig. 1), was made at home from a ten-cent collapsible crate. Any one can make it. This tray holds 13 pint jars; or 10 quart jars; or seven 2-quart jars. Individual jar holders made of wire may be purchased. (See Fig. 2.) These may be obtained for 10 cents each from your local dealer, or from the Handy Manufacturing Company, Seattle, Wash.

Fig. 2 — Handy Jar Holder — Observe the Projection on the Bail, Causing It to Stand Up for Easy Handling

17

SOME POINTS TO REMEMBER

Anyone can do Cold Pack Canning with any style of glass jars or tin cans which can be used for hot-pack canning (except the wax-sealed cans), and in any boiler or kettle that is deep enough. You do not have to buy a special canning outfit.

But do not forget these points:

1st — Have a false bottom that keeps the jars at least three-fourths of an inch from the bottom of the cooker.

2nd — Have the false bottom slatted or open so that the water will circulate around the jars.

3rd — Have the water come an inch above the top of the jars.

Lifting Tray of Canned Vegetables from Cooker. Home-made Tray

4th — Do not begin to count time until the water is boiling. Water is not boiling when small bubbles appear on the bottom of the kettle, nor even when they form all around the sides and rise to the top of the water. It must bubble hard all over the top.

5th — Keep at a lively boil until the time is up.

6th — Do not let the jars cool in the water, when the cooking is finished.

Anyone can can any food product, fruits, vegetables, meats, fruit juices, greens, soups, fish, game or fowl by the Cold Pack method.

TYPES OF FACTORY MADE OUTFITS

When canning is done on a large scale it is advisable to use a factory-made outfit; it saves time.

There are several good types of outfits, and the profits from one seasons's work will more than pay for one.

Most firms which make canning outfits manufacture some styles which are self-heating; and for a little more money, a canner with its own firebox, which can be used out of doors or in a special room may be obtained. These are especially desirable for community and canning club work.

In **Steam Outfits** the jars do not sit in water, but in a tray or crate above the water. A small amount of water in the cooker forms steam in which the products are sterilized.

A High-Pressure, All-Aluminum Steam Cooker is especially desirable for use in high altitudes, and for products such as corn, pumpkin, etc., which require a high temperature or long cooking.

Steam Canner

Products will cook in such a cooker in one-third the time required to cook them with an ordinary hot-water outfit; in some cases the saving of time is even much greater than this. The all-aluminum boiler can be subjected to intense heat and pressure.

Pressure Cookers are much used west of the Rockies for preparing meals and are rapidly coming into favor in the east because of the short time required for cooking foods.

Steam Canners are of aluminum, steel, iron, or boiler plate, and will not

Pressure Cooker

19

stand quite so high a temperature as the pressure cookers.

Safety Steam Cooker—The Special feature of this Cooker is that the top cannot be unclamped or opened in any way while any pressure remains in the kettle. This prevents accidents from escaping steam. The operator must first open the blow-off valve at the top of the kettle and allow the steam to escape in that way.

This safety provision is an especial advantage with young people or those not accustomed to working with steam-pressure outfits.

Water-Seal Steam Canners give a temperature slightly above boiling point. These are of galvanized iron.

The cylindrical cover sets into a double jacket, and the extra air-space helps maintain a temperature of 2° above the boiling point. (In the altitude of

The Sprague Safety Cooker

Chicago, about 600 ft., water boils between 210° and 211°.)

A Rapid-heating Firebox is a feature of the outfit shown on Page 21, which illustrates this type.

The Cover, Kettle, Jacket, Heater, and Wire Crate of the Safety Cooker

Hot Water Bath Outfits may be of tin, copper, iron, or galvanized iron. In the hot water bath outfit, enough water is put in the cooker to extend one inch above the top of the jars, and the goods are cooked at the boiling point.

The home-made outfit is a hot water bath outfit. The commercial hot water bath outfits are similar to a large kettle except that they are manufactured for use in canning and so are suited in size to hold jars economically.

Time-table — The time-table for hot water bath outfits is based on quart jars, cooked at 212°. In high altitudes water boils at lower temperature and so it is necessary to cook products longer.

For cooking in a *hot water bath outfit,* the time must be increased as follows:

500 to 1500 feet, use time-table as given

1500 to 3000 feet, add 10 per cent

3000 to 4000 feet, add 20 per cent

4000 to 7000 feet, add 40 per cent

The time-table for *steam* outfits does not vary, but is the same for all altitudes.

Water-Seal Steam Outfit
with a Rapid Heating
Firebox

Courtesy Home Canner Co., Hickory, N. C.
Group of Girls Using an Out-Door Commercial Hot Water Bath Outfit

PRESERVE YOUR HOME 139

OTHER THINGS NEEDED

In addition to a canning outfit and cans, we need sundry other articles.

Tables — If several are to help in canning, be sure to have plenty of table space — one long table or several smaller ones set end to end.

Chairs — Plenty of them, so the workers will not tire at their work.

Pail for Blanching—It saves time to have a separate pail, pan, or kettle for blanching. In this way products for the second pack can be blanched while the first are cooking. Add a cover to hold heat and a wire basket for blanching in steam.

Covered Steamer Set in a Pail Makes a Good Outfit for Blanching in Steam

We shall also need a piece of common cheesecloth, a towel, or a wire basket, in which to put the vegetables for lowering them into the hot water.

The Wire Jar Holder (See cut Page 17), **a Duplex Fork, or a Wire Potato Masher** of the type shown in lower cut on Page 29, may be used for lifting single jars above the water so they can be lifted out with the towel.

The Dish Pan May Be Used for the Cold Dip

Pails, Pans, Basins, Sharp Knives, Spoons, a Measuring Cup, Can Filler, Colander, etc., for use in preparing the product; a **Clock** to time the processing; **Towels, Labels, Paste,** and **Brush** for labeling cans, if we are canning to sell; **Scales** to weigh the filled cans and see that each is standard weight; in brief, such utensils and supplies as are necessary to quick, accurate work, should be provided.

Keep a Note Book of Information about the variety and state of the product, time blanched, grade of syrup used, time cooked, and any special features — facts which might affect quality or keeping.

22

If tomatoes canned after such and such a date, or blanched or cooked too long, or too short, a time, are not as high grade as those canned under other circumstances, that is a good thing to know so our next year's pack may be improved.

Date All Goods — It helps locate them and is an index as to when to dispose of such goods as are best within a limited period after canning.

STEPS IN COLD PACK CANNING
No Preservative Needed

It is quite unnecessary to use any canning compound or other preservative. Cooking the product in a closed jar according to the instructions given, will sterilize any food so that it will keep without a preservative.

In learning to can by the Cold Pack method, it is well to begin with one product only, and with only a small quantity. Then we are not hampered by too many things to do all at once and can familiarize ourselves absolutely with every step. When we feel at home in the work, then we can undertake larger quantities and new varieties.

STEPS IN COLD PACK CANNING

NO PRESERVATIVE NEEDED

SELECT SOUND PRODUCTS
GRADE WASH TRIM
SCALD OR BLANCH
COLD DIP QUICKLY
PACK CAREFULLY AND CLOSELY

ADD HOT WATER OR SYRUP
PLACE RUBBER AND COVER ON JAR
DO NOT SEAL GLASS JARS TIGHT
COOK PER TIME-TABLE

HOT FIRE PLENTY OF WATER
THINGS HANDY

CHART V

The eight photographs shown on pages 24 and 25 indicate plainly each step necessary in the canning of beans by the Cold Pack method.

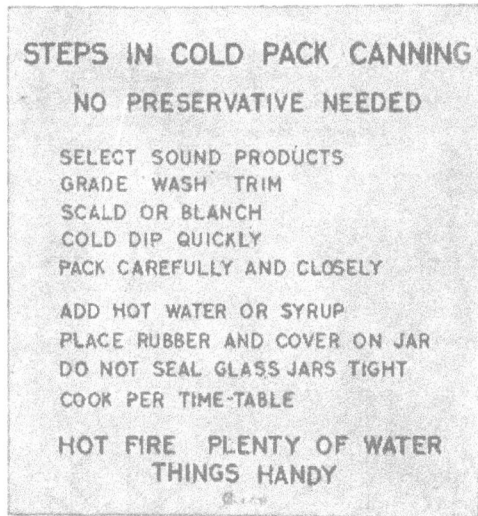

Don't have an empty preserving jar in your home next fall.

STEPS IN CANNING *

1—Preparing the product

2—Scald or blanch

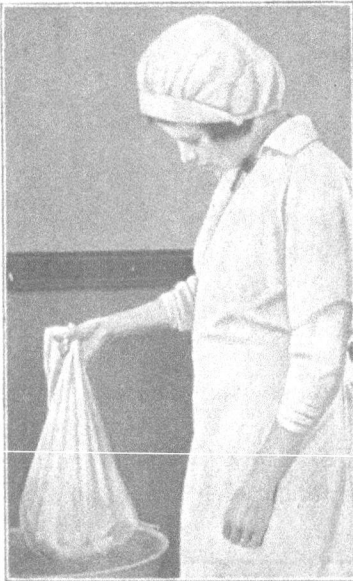

3—Plunge into cold water

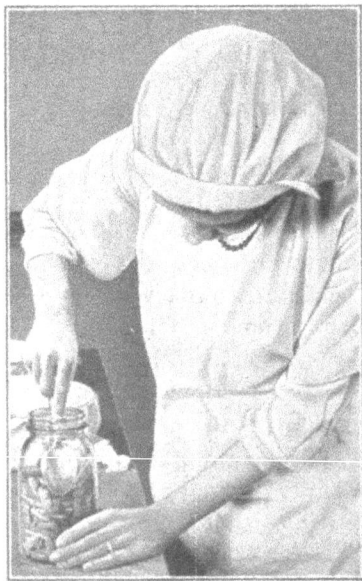

4—Pack close

* Beans are here used as an illustration for canning vegetables.

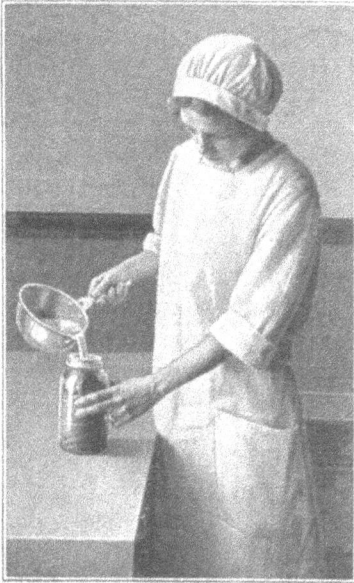

5 — For vegetables add salt and hot water (if fruit, add syrup only)

6 — Put on cover — not tight

7 — Place in cooker — cook per time table (page 60)

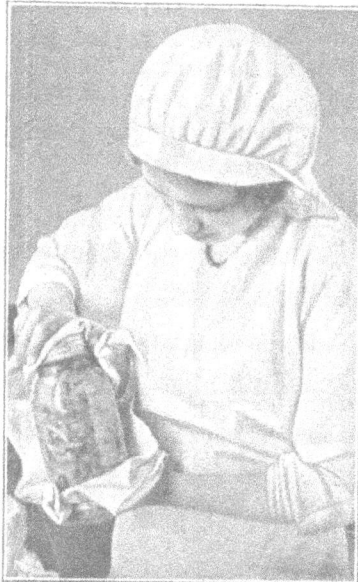

8 — Seal tight at once

PRESERVE YOUR HOME   143

CANNING TOMATOES

Select Sound Products — Select fresh, ripe, firm tomatoes.

Grade, Wash, Trim — Grade for ripeness, size, and quality; this is to insure a high-grade product. Can different sizes and shades together. If the products are of the same ripeness and quality, they will cook better, and be more attractive in appearance.

Of course, we wash the products clean, and where necessary trim them—pare the apples, string the beans, silk the corn, hull the berries — in short, prepare them as may be necessary. (In the case of tomatoes, remove the skin and the stem end *after* scalding.)

Scald or Blanch—Scald means to immerse in boiling water. Blanching is a longer process. Both loosen the skin. Blanching will also reduce the bulk, and drive out objectionable acids.

Tomatoes need to be scalded only enough to loosen the skin.

Have ready a kettle of boiling water. Put the tomatoes in a wire basket, or, lay them on a piece of cheese-cloth, or a towel, twist the ends together to form a sack, and let this down into the kettle. It is a good plan to slip a rubber band around the neck of this sack to hold the ends in place. The ends should be long enough to stand up out of the water and so avoid danger of burning the fingers when removing the product.

Have the water boiling hard and leave it over the fire so they will scald quickly. If the water is not boiling it is difficult to loosen the skins without leaving the tomatoes in so long that the pulp becomes soft.

If the tomatoes are ripe and the water is boiling, one-half minute to one minute will be sufficient; unripe tomatoes may require longer. A little experience will enable us to tell just when they are scalded enough.

Do not leave the tomatoes in the hot water until the skins break, as this gives them a fuzzy appearance.

Cold Dip Quickly—Lift the tomatoes out of the hot water and plunge them immediately into cold water.

26

The cold dip makes them easier to handle, separates the skin from the pulp, firms the texture, and coagulates the coloring matter so it stays near the surface, giving them a rich, red color.

Then the shock due to the sudden change from hot to cold and back to hot again seems to help kill any bacteria or organisms.

Do not let the product stand in the cold water. The water becomes lukewarm, softens the product and allows bacteria to develop.

Trim and Peel Tomatoes—Take the tomato in the left hand and with a sharp knife cut out the stem end. Be careful not to cut into the seed cells, or the seeds and pulp will be scattered through the liquid, injuring the appearance of the product. Cut out the stem end and then remove the skin.

Pack Carefully and Close — The jars, rubbers, and tops, should be ready. Glass jars should be hot, so there will be no danger of breakage in setting them into the hot water, and so they will not cool the water in the cooker below the boiling point.

Pack the tomatoes whole, pressing and shaking them well down together, but not using force enough to crush them.

Add Hot Water or Syrup— The instructions on the chart are general—hot water for vegetables, hot water or syrup for fruits.

Tomatoes are an exception; as a large part of the tomato is water, no liquid is needed.

Under the Pure Food law canned tomatoes to which water or extra juice has been added are con-

Pack the Product Close

sidered adulterated. This is intended to protect the public against unscrupulous canners who might slack fill the can with tomatoes and add surplus water or juice. If the tomatoes are to be sold, therefore, nothing should be added except one teaspoon of salt, or a teaspoon full of salt and a half spoon of sugar per quart, for seasoning.

If the tomatoes are to be used at home they may simply be packed close together, or, those which break in handling may be crushed and poured over the whole tomatoes to fill the spaces. Where tomatoes are to be used as stewed tomatoes, or for soups, they may be cut in pieces, as more can be packed in the same space than if packed whole, but do not fail to can some whole for salads and exhibits.

Jars Do Not Need to Be Full in Order to Keep. By the hot pack method the air in a jar which is only part full has not been sterilized, and may contain bacteria which cause the product to ferment or mold. But by the Cold Pack, the air in the jar is sterilized while the product is being sterilized, and if the jar is closed before cooking, a single spoonful may be canned in a two-quart jar and the product will keep as well as though the jar was full.

Using Thumb and Little Finger to Screw Down the Cover

Place Rubber and Cover on Jar — Fit the rubber. Use good rubbers and see that they lie flat and fit close up to the jar. Put the covers in place.

Do Not Seal Glass Jars Tight — If using screw top jars, screw the cover down until it catches, then turn a quarter of an inch back; or screw down with the thumb and little finger, not using force, but stopping when the cover catches.

If using vacuum seal jars, put the cover on and the spring in place. The spring will give enough to allow the steam to escape.

In using glass top jars with the patent wire snap, put the cover in place, the wire over the top, and *leave the clamp up.*

The cover on a glass jar must not be tight while the product is cooking, because the air will expand when heated, and if the cover is not loose enough to allow the steam to escape, the pressure may blow the rubber out, or break the jar.

Set in Cooker — After the products are packed, it is an

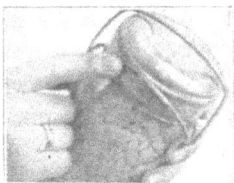

Leave the Clamp Up

advantage to cook as quickly as possible. Time lost in bringing the contents to the point of sterilization softens the product and results in inferior goods.

For most products, we pack in hot jars, fill with hot water or syrup, have the water in the cooker boiling and have enough water so it will not stop boiling when the jars are set in. If we use ordinary good sense in handling the jars, we will have no breakage. But tomatoes are only slightly warmed in blanching, and as we add no hot water, the jar is not hot enough to make it safe to set it directly into boiling water. *Jars of tomatoes should be set in warm water until ready to place in the canner.*

Cook Per Time-Table — If products are undercooked they will not keep; if they are overcooked they lose flavor and texture.

Tomatoes sterilized under boiling water require twenty-two minutes. Berries, apples, and small fruits, will process in five to twenty minutes; greens require twenty-five minutes to two hours; and sweet corn, forty-five minutes to three hours, according to the outfit. (See time-table on Page 60.)

Do not begin to count time until the water is boiling.

Hot Fire, Plenty of Wa-ter, Things Handy — We

Using a Wire Potato Masher to Lift Jar From Cooker

must be able to secure a hot fire quickly, and should keep a fairly even heat. Do not try to economize on water. We must have plenty of clean water to wash jars and products, to make syrups and brines, for use in blanching, and, if we are using a hot-water outfit, for use in processing.

FINISHING THE WORK

Remove Jars From the Cooker—Do Not Expose to Cold Drafts—In removing canned goods from boiling hot water, protect from drafts. Do not set in an open window; drafts might break the jars.

FINISHING THE WORK

REMOVE JARS FROM COOKER
DO NOT EXPOSE TO COLD DRAFTS

EXAMINE RUBBERS TIGHTEN COVERS
INVERT TO TEST THE JOINT AND COOL

WRAP TO KEEP OUT LIGHT
STORE IN COOL DRY PLACE

STICK TO ONE SET OF INSTRUCTIONS
WORK QUICKLY
HAVE EVERYTHING CLEAN AND SANITARY

ATTENTION TO LITTLE THINGS
PRODUCES HIGH GRADE GOODS

CHART VI

Examine Rubbers. Tighten Covers— Examine rubbers to see that they are in place.

Sometimes if the covers are screwed down too tight, the pressure of the steam from the inside causes the rubber to bulge out. Simply loosen the cover a thread or two and push the rubber back into place and then tighten. In case the rubber does not seem to fit well, or seems to be a poor rubber, it should be replaced by a new one and the jar returned to the cooker for five minutes.

The jars should be sealed tight—covers screwed down, clamps put in place—immediately after they are removed from the cooker.

Invert to Test the Joint and Cool—If the seal is not perfect, correct the fault, and return the jar to the cooker for five minutes if hot, ten minutes if jar is cold.

Do Not Invert Vacuum Seal Jars. These should be allowed to cool and then be tested by removing the spring or clamp, and lifting the jars by the cover only. Lift the jar only a half inch, holding it over the table so that in case the lid does not hold, the jar and contents will not be damaged. Or, better still, tap around the edge of the cover with a rule. An imperfect seal will cause a hollow sound.

30

Wrap to Keep Out Light. Store in a Cool, Dry Place— Light injures some canned goods; bacteria breed in heat; dampness favors mould and may cause rust. Canned goods are best kept at a temperature below 70° F.

Canned goods exposed to very unfavorable conditions may lose the delicate flavor and color, and in some cases may even spoil.

Do not spend your time canning fruits or vegetables and then allow them to spoil because of improper handling afterward. Do not condemn factory canned goods which have been stored in a hot room.

Stick to One Set of Instructions—If you have several different sets of instructions you may be interested to try out each of them and see which is the most efficient, the least labor, and produces the most satisfactory results, but *do not combine* two sets of instructions—you will be certain to get into difficulty.

Work Quickly—Take the steps in rapid succession: The cold dip should follow the blanch immediately—the product should be packed and hot water or syrup added at once—it should be processed as soon as possible after packing; else the beneficial effects of shock on the bacteria will be neutralized. All along the line, quick work is an advantage, is safer, and produces better results. If we are to can in quantities we must work quickly and surely, else our profits will vanish.

Have Everything Clean and Sanitary—Absolute cleanliness is necessary. Dress, Hands, Jars, Tables, Utensils, everything used about the work, should be absolutely sanitary — sterilized where necessary and scoured clean always.

O. H. Benson and Geo. E. Farrell of the Office of Extension Work, North and West, Washington, D. C., Training Teachers in the Canning Work at the Office of Co. Supt. E. J. Tobin, Chicago

Use hot soapsuds freely for cleansing utensils. Especially do not use jars or covers which cannot be cleaned perfectly.

Wash the products in pure water. Impure water is offensive and may spread disease. Scrub the products with a brush if necessary, and rinse thoroughly through several waters.

If any product cannot be made perfectly clean, do not use it.

We are preparing food to be eaten and must comply strictly with all sanitary requirements. To take chances is to endanger the health of the consumer. The room must be screened, the hair may be protected by a cap, and the dress by a clean apron. Be sure the hands are clean.

In high-grade commercial canning factories much attention is given to screening, light, ventilation, drainage, paint, and white-wash. Personal cleanliness is enforced.

Attention to Little Things Produces High-Grade Goods— Perfection of detail makes the difference between fair and excellent.

If we have no other reason for canning, let us can and sell, and with the money buy modern conveniences for the home. Any home can have a better water supply, better lights, labor-saving machines, such as a vacuum cleaner and a canning outfit; we can can and earn the necessary amount which we would not otherwise have had.

It is not always having so much to do which makes women's work hard; it is more often having to do something and having neither the supplies nor the utensils to work with.

Courtesy of Co. Supt. E. J Tobin

A National Leader, a State Leader, a School Teacher, and a Rural Canning Club at Work

IT'S GOOD BUSINESS

An Average Profit, 1/10 Acre Tomatoes

This account is an average made up from the records of Tomato Club girls. The rent is figured at $1 per tenth acre, and the labor (Club girls') at 10c per hour. (Read chart.)

The average profit reported by Canning Club girls in 1916, from a tenth of an acre was $24, or at the rate of $240 per acre. Some of the Club girls have made more than $100 from their tenth acre.

(The lecturer may give the year's best record among Canning Club girls, the record for the county, the home state, etc., if the figures are

IT'S GOOD BUSINESS

AN AVERAGE PROFIT 1/10 A. TOMATOES

RECEIPTS

SOLD FRESH	$10.40
SOLD CANNED	18.40
USED AT HOME	9.60
ON HAND	2.95
TOTAL	$41.35

EXPENSE

RENT FERTILIZER PLOWING	2.50
CULT'VT'G STAK'G PRUN'G	3.75
HARVESTING AND CANNING	5.25
CANS JARS LABELS	5.85
TOTAL	$17.35
NET PROFIT	$24.00

AVERAGE COST PER QT. 4¢

CHART VII

available.) The average cost per quart of home-canned tomatoes is 4 cents.

Marketing Canned Foods

In most cases, if our goods are of standard quality and pack, and we can guarantee a definite number of cans, the home grocer will buy them as readily as he will jobber's goods. He must be able to depend on us, and we must make arrangements some time in advance, as grocers usually place their orders early in the year.

Sometimes we can't sell garden produce fresh, because when we have tomatoes or beans, everyone else has all he can use. But if we can them and put them on the shelf until the fresh vegetables and fruits are gone, then people will be glad to pay a good price for our canned foods.

With a few exceptions, such as apples and blackberries, canned goods do not deteriorate but will keep indefinitely, and so may be held until the market is favorable.

33

WE GROW IT, WHY NOT CAN IT?

Red Tomato on his way from the garden to the winter dinner table. The grower may sell him to the commercial canner (at $8 to $10 per ton), from the cannery the canned goods may pass to the wholesaler, who sells them to the retailer. The housewife may buy them at the country store at 15c per can or at a price of $120 per ton.

CHART VIII

There is room for the commercial cannery just as there is room for the commercial bakery, or laundry, or tailor shop, but let us not be dependent on the commercial canner. A can of corn may be **Grown** in the middle West, **Canned** at a commercial cannery, **Shipped** from there to a Baltimore jobber, **Sold** to an Eastern wholesaler, **Then** to a Chicago commission house, **Next** to a middle West retailer, adding **Profit** and **Transportation** charges all along the way to be **Bought** by folks who fed bushels of sweet corn to the hogs, because it reached the eating stage faster than they could consume it.

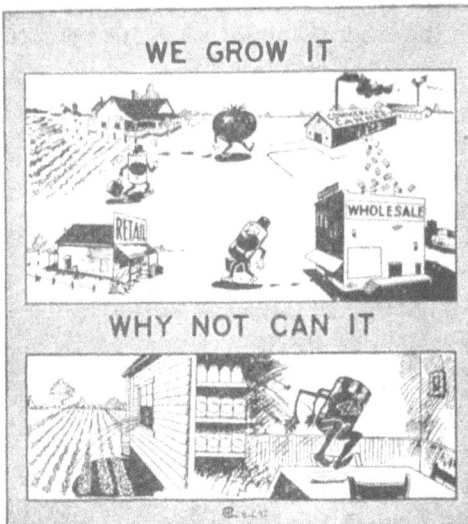

Home Grown, Home Canned

How much better it would be to can that corn fresh from the home field and store it on the shelf for winter use. **Grown at Home, Canned at Home, Used at Home.**

34

CLUB WORK GIVES 4-H TRAINING

O. H. Benson says that Club work is "the right hand of fellowship from the Home to the School and from the School to the Home."

Modern understanding of education is that it is a training for citizenship, and that such training should not be one-sided, but should train:

The **Head**—To Think, to Plan, to Reason.

The **Heart**—To be Kind, True, and Sympathetic.

The **Hands**—To be Useful, Helpful, Skillful.

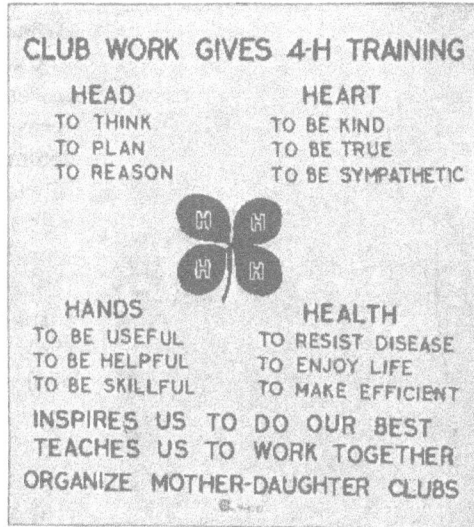

CLUB WORK GIVES 4-H TRAINING

HEAD
TO THINK
TO PLAN
TO REASON

HEART
TO BE KIND
TO BE TRUE
TO BE SYMPATHETIC

HANDS
TO BE USEFUL
TO BE HELPFUL
TO BE SKILLFUL

HEALTH
TO RESIST DISEASE
TO ENJOY LIFE
TO MAKE EFFICIENT

INSPIRES US TO DO OUR BEST
TEACHES US TO WORK TOGETHER
ORGANIZE MOTHER-DAUGHTER CLUBS

CHART IX

The **Health**—To Resist Disease, Enjoy Life, Make Efficient.

The four-leaved clover, a leaf for each H, is the emblem of the Boys' and Girls' Clubs. The Canning Club motto is:

"Make the Best Better."

Inspires Us to Do Our Best—We do more work and better work when we work together. We each want to do as well as our neighbor, and we put into our work the best we have. Then we exchange ideas and help one another. It is no longer "I" or "you," nor even "us;" it is "our neighborhood"—and we all put forth our best effort, and all pull together for the honor and development of the community.

Teaches Us to Work Together—Canning in clubs makes for neighborliness. We forget petty jealousies. We come to understand the neighbor we did not like and decide that she isn't such a bad sort after all. We do better work. Ten girls working together will do more canning and better canning than

35

ten girls working separately. We exchange suggestions and experiences and all profit by the increased knowledge.

O. H. Benson

Working With Others Broadens Ideas and Ideals — Canning in clubs is certain to lead toward a general interest, both in canning and in community needs and possibilities. It establishes connection with the outside world in at least three directions: with the state schools through the State Canning Club Leader, who is associated with the university or agricultural college; with the U. S. Department of Agriculture, which supervises the club work; and with the business world, such as local grocers, commission merchants in large cities, hotels, restaurants, hospitals, railroad dining car service, and other possible markets. The boys and girls learn to think in terms of these larger units and to plan their work to meet the new conditions.

The Home Canning Club meets with the canning club from some other neighborhood and all of them keep in touch with Canning Clubs in other counties and so in touch with the world. All may come together at the State Fair and consult with the State Leader of Boys' and Girls' Clubs and with the Specialist in Charge of Club Work at Washington. All in all, it is a great social influence.

We not only work to better advantage when working together, but the association is an inspiration. Most communities have

Courtesy of Home Canner Co.

Group of Southern State Leaders

possibilities.   They could do more if only there were "someone to go ahead."   Club work, community work, a common interest, develops leaders—and when I say "leaders " I do not mean those who **stand** and **say,** **"Go;"** but those who **go** and **do,** and **say,** **"Come."**   We need more leaders who lead by serving.

Organize Mother-Daughter Home Canning Clubs—Boys' Clubs are good, Girls' Clubs are good, Men's organizations, Women's organizations—there is a place for each of them, but there is a new club, the Mother-Daughter Club.

It has the usual club possibilities for good times, it spreads useful knowledge, and it gives Mother and Daughter a common work and pleasure.   Then when Mother and Sister work together, Brother is anxious to help pick and peel and solder and keep the 'fire going; Father likes the business-like air there is to the new way, and after he makes a few calculations he realizes that, whether you figure it by the day or by the acre, the "women folks" are making about as good money as there is in corn or cotton, and Father becomes interested and helps can and tip and lift and carry.

No, we don't want to add to women's work; we want to lighten it.   It is very much easier to get a meal when we have canned in a business way and in sufficient quantity.   Then we have on hand a variety of good healthful foods and we do not have to wonder what to cook.   It is very much easier to cook twenty-five cans of tomatoes at one process than it is to cook twenty-five individual lots.

Glenwood, Neb., Mother-Daughter Team Canning at Home

WHY HAVE I BEEN TALKING TO YOU ABOUT HOME CANNING BY THE COLD PACK METHOD?

WHY HAVE I BEEN TALKING
TO YOU ABOUT HOME CANNING

WASTE! WASTE!! WASTE!!!
WHAT IS WASTED WOULD FEED US
ANYTHING CAN BE CANNED AT
  HOME BY THE COLD PACK METHOD
GIVES US THE RIGHT KINDS
                    OF FOOD
IT'S GOOD BUSINESS IT PAYS
YOU'LL LIKE IT BETTER
YOU 'GREW IT YOURSELF
YOU'VE HELPED MAKE
        A BETTER NEIGHBORHOOD

IF WE DON'T CAN IT
  WE WON'T HAVE IT

CHART X

Because of **Waste! Waste!! Waste!!!**

What Is Wasted Would Feed Us— When there is so much want all about us, it is shiftless and wicked to let any good food go to waste.

Anything Can Be Canned at Home by the Cold Pack Method— Save the waste and feed the hungry.

Gives Us the Right Kinds of Food —Juicy, nutritious, palatable vegetables and fruits which our bodies need. And besides,

It's Good Business. It Pays—It reduces the cost of living and it may be made a source of income.

You'll Like It Better. You Grew It Yourself—There may be a difference of opinion as to whether home-canned or commercial-canned tomatoes are best. There is no question as to which is better, but, our own has an extra flavor—born of the pride which comes with owning something—from doing the work ourselves. There is no prouder moment in our lives than when we proffer another the first fruits of our own labor.

"We grew these strawberries, Mrs. Brown. Yes, right in our own garden. Johnny made the beds, set out the plants, and weeded them himself, and he's proud as a hen with one chicken. Brings all the boys in the neighborhood in to admire 'em an' begs me to make shortcakes for 'em. Yes, it's some work, but I'd rather have 'em here than off by themselves learning bad habits. He's got so he takes an interest."

38

Or, we pass the preserves with an extra pride. "Mary made them. Did it in Club. They're quite the finest we ever tasted. Yes, that new teacher does contrive to keep the children interested. Always got something new. Get their lessons, too. Don't seem to interfere with their school work."

You've Made a Better Neighborhood—The Canning Club should not be a temporary organization for the sake of "doing something new." It should aim at permanent results in canning; improved methods, and a more general canning of vegetables, fruits, fruit juices, soups, and meat. The study of Home Economics should develop higher standards of sanitation and of general living conditions. Good taste in dress, furniture, landscape gardening, and in standards of conduct will all grow out of a properly conducted club under a competent, devoted leader. Club canning makes for neighborliness. Everybody helps everybody else. We study and plan together for a better community. In the end it is citizens and home makers we are making, not simply canners.

If We Don't Can It We Won't Have It—Some people may say that they would rather grow corn or cotton and buy their canned goods. The answer to this is that they do not buy enough. The final argument for most of us is: "If we don't can it, we go without it."

Organize—Shall we consider the appointment of a committee to arrange date and place of meeting to organize a Home Canning Club? Or, shall we organize now?

Well Put Up Exhibit of Home-Canned Foods Packed by Four Sixteen-Year-Old Girls, F. L. Audrian, Supt. City Schools, Kiona, Wash., Club Leader

Opportunity For Boys and Girls

Although jobbers buy commercial canned tomatoes at 75 cents a dozen, the average retail price is 15 cents per quart, and the average cost of canning tomatoes at home or in club is 4 cents per can. What is the average price of tomatoes at your grocery store? What is the saving per can?

Suppose a girl cans 100 quarts per day—what kind of wages can she make?

Suppose we figure on the club slogan "a can of fruit, a can of greens, and a can of vegetables for every day in the year"—approximately 1,100 cans at an average saving of 10 cents a can, $110; isn't this worth saving?

And at that we have left out the soups and fruit juices, both of which are easy to can and desirable to have.

Here is the finest opportunity in the world for a girl to earn money at home. She can be of service, save waste, earn money to go to school, etc. One of the first great steps in citizenship is helping one another to be of service.

Besides buying her own clothes, helping out the family, and gratifying some special wants, a Canning Club girl can start a bank account and save enough money to go away to school.

Besides that, she has a profitable business all her own. She can earn much more than by clerking in a store, or working

Courtesy Home Canner Co.

Capped and Aproned Club of Girls at Work

for $8 to $12 a week as a stenographer. If she is enterprising and a good business woman she may make more than by teaching school.

Cold Pack canning produces a commercial product worth good money. If all people in the community who have canned goods for sale club together and work together it is easier to secure a market. Then, too, the girl who cans is learning how to do the work which is distinctly her work. Generally speaking, the feeding of people is the woman's business.

The Mothers' Responsibility

The women of America owe a duty to the country in the present crisis. We should preserve all surplus food. America will need it.

"Every morning the world wakes up hungry," and every day the women of the country must relieve that hunger. Some of them do it haphazard way, just any old way and anything; some of them study Foods, and Health, and Hygiene, and plan their menus; some of them realize that not only the health of the world but the business of the world, is dependent on the breakfasts and dinners and suppers that the world eats.

The *Health* of the human race is in the hands of the women of the world.

The child who eats the foods which make bone and muscle and vitality can grow bones and muscles and good red blood. Stunted, pale, sickly, weak children often indicate children not properly fed.

Boys and girls who get plenty of fresh air and exercise and who eat properly will outgrow many hereditary weaknesses. Air and exercise are the property of him who will take them, but we must eat what is set before us.

We Balance Rations for the Stock, Why Not for the Home Folks? We are so intent on the moral and mental welfare of mankind, that often the physical is neglected. We know that animals must have a balanced ration, and we send our sons to agricultural colleges to learn about suitable feeds for stock; we must also insist that not only colleges and high schools, but every one-room school as well, should teach both girls and boys to regulate their eating with a view: **1.** To repair waste, **2.** Maintain health, and **3.** Furnish energy.

Planning and constructing a healthful, satisfying, tasty dinner and setting it before the family in a dainty, artistic way is as fascinating and quite as useful as designing and making a hat, or painting china. And—don't forget this—if we make it so by painstaking, competent service, quite as dignified and honorable.

The *Health,* the *Business* and the *Happiness* of the world is in the hands of women.

People cannot be alert, clear-thinking, clean-acting, and efficient unless they are fed properly.

The *Happiness* of the world is in the hands of the women, because poorly fed people

> Become ill-tempered—and quarrel;
> Become dissatisfied—and indulge in drink, questionable amusements, and bad company;
> Become sick—and perhaps lose health permanently;
> Become discouraged—and quit.

Quit work, quit home, quit morality and manhood and character, quit trying,—just quit.   And when a man, or woman, quits, unless we can get them back mighty quick, the game is ended.   There is nothing more to be said or done.

Community Life, Health, Business Efficiency, Happiness— it is a large order, but it hinges absolutely and undeniably on the diet; and in the country, at least, the diet hinges partly on Home Canning in club work.

Courtesy of U. S. Dept. of Agriculture
Laying the Foundation of Community Life, Health, Business Efficiency, and Happiness

CANNING IN TIN

People who can in large amounts usually can in tin for the following reasons:

Shipment Is Easier—People who can in large quantities can to sell, and there is danger of breakage when products are shipped in glass jars. Tin cans are lighter weight; they require less space; the cans need not be returned.

Tin Cans Are Less Expensive—If we sell our canned goods, tin cans are less expensive. For home use, glass jars are cheaper in the end, as they can be used several years in succession, while tin cans must be replaced each year; but if we are canning to sell, it would be necessary to add the cost of the jar to the cost of the product, or, to require the glass jars returned to us for use the next year.

There Is No Danger of Breakage; Less Storage Space Is Required—The tin cans can be handled more readily—they may be set under the faucet or in a vat for rapid cooling, and they may be stacked one on top of another. This last makes it possible to store them in much less space than is required for glass jars.

Caution—In handling the ''sanitary'' can before packing, care must be taken to see that the flange at the top is not broken or cracked, or it will be impossible to make a perfect seal.

The Product Sells More Readily—Except for fancy trade, few commercial goods are canned in glass. In addition to tin canned goods being cheaper (we have to charge a higher price to cover the cost of glass cans), there is the fact that the general public may can at home in glass, but it is accustomed to buying its canned foods in tin, and custom is a big factor in business.

Lacquered Cans Should Be Used for Very Acid Products — Such as red fruits — pumpkin, squash, sweet potatoes, red beets, sour cherries, gooseberries, blackberries.

Concentrate products so that each jar or can will hold as much food and as little water as possible.

43

SEALING TIN CANS

The Hand Sealer

Simple, inexpensive, hand-sealing devices for sealing the "sanitary" can without the use of solder, acid, or heat are now on the market which no doubt will soon be used extensively.

Simple Hand Sealer

The entire top of the can is open, which makes it possible to pack tomatoes peaches, etc., whole. The work of sealing is simple.

The product is packed in the can, the cover placed over the top and then inserted in the sealer. By turning the crank a few times the seam is rolled down, making an air-tight seal. This machine can be obtained for about $10, not including cost of cans.

Full information can be obtained by writing the manufacturers, Burpee and Letson, Ltd., South Bellingham, Wash.

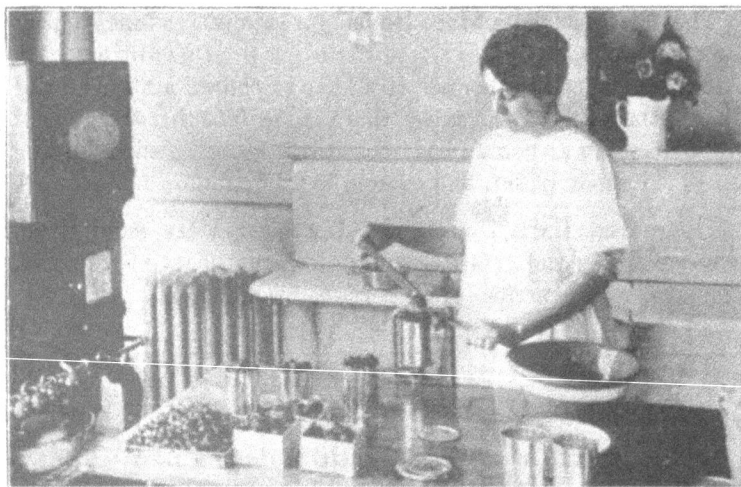

Using the Hand Sealer — Notice the Open-Top Cans

44

STERILIZING PRODUCTS IN TIN CANS

It Is Not Necessary to Exhaust — Some canners cap the cans and then exhaust before tipping. This method requires an extra handling of the cans, and is not necessary where the product is *blanched* and *cold-dipped* before packing, and *hot* water or syrup used to fill the can.

Intermittent processing requires so much extra time and so much unnecessary lifting that it kills enthusiasm. Where time, labor, and fuel are valued, it is quite as cheap to buy vegetables ready canned, as it is to can by the intermittent process, and most people will prefer to do so.

Probably it is wise to follow the instructions sent with our canning outfit, or given us by our state club leader, until we are sure of our method; then, if we wish we may try out other methods and choose the one we like best.

There is no danger of breaking tin cans but there is danger of over-heating the product in the bottom of the can, so we need a tray, or the wire basket, the same as for glass jars.

Leaks—If a can leaks, air bubbles will rise from it when it is set into hot water. The cans should be turned upside down to discover any leaks which may have escaped notice in cooking.

Cool Quickly—If tin cans are packed close together when they are taken from the cooker, and allowed to stand, there is danger that the cooking process will continue and the flavor and color be injured. They should be placed immediately in a bath of cold water or under the cold water faucet.

Mark the Cans So You Can Distinguish Them—With a rubber stamp or pencil mark the cans before putting them in the cooker. This is the only safe, sure way to keep from getting the different products mixed.

When the cans are perfectly cool we can set products of a kind in one place, those of another kind in another place, and so distinguish them until they are labeled, but while we are handling them, care is necessary to keep from getting them mixed.

Produce in your garden lots of cabbage, potatoes and root crops that can be kept over the winter without canning.

45

LABELS

Label the cans when the product is ready to sell; then the labels will be clean and attractive for the customer.

Small labels such as we use on glass jars are not advisable for tin. With these it is necessary to roughen the tin with the acid at the point where the paste is to be applied. Even then, if the can is set in a very dry, warm place, the label may drop off. If set in a damp place, the can is apt to rust where the paste was applied.

The type of label, used by commercial canners which is placed around the entire can and is fastened with paste applied at the ends only, is the best style. This is similar to the label shown below.

The regular club label which carries the 4-H brand, the club motto, and blanks for weight of contents (without juice), date canned, and name and address of canner, is neater and more suitable than pictures of inappropriate flowers or pretty girls. Print at least one recipe on each label.

Buying Food to Can

One big important feature of Home Canning is to save the products of the orchard and garden which are now allowed to go to waste; home canning will reduce expenses even when the food to can must be bought on the market.

In 1916, the Uncle Sam, Preparedness, Ever Ready, B. and G., Happy Helpers, and Economy Canning Clubs of the public schools of Pawtucket, R. I., working under the direction of Miss Alice L. Currier, Supervisor, bought fruits and vegetables on the market and canned the following: 7½ qts. apples, 23½ qts. blackberries, 41 qts. blueberries, 16½ qts. currants, 6 qts. gooseberries, 71½ qts. peaches, 20 qts. pears, 13½ qts. pineapple, 93 qts. raspberries, 1½ pts. rhubarb, 6½ qts. strawberries, 3½ qts. asparagus, 51 qts. snap beans, 49½ qts. shell beans, 45 qts. beets, 13 qts. beet greens, 30 qts. peas, 68 pts. peppers, 25¼ qts. summer squash, 5 qts. sweet potatoes, 4½ qts. spinach, and 17 qts. tomatoes, a total of 577 qts. **Investment, $137; net profit, $249.**

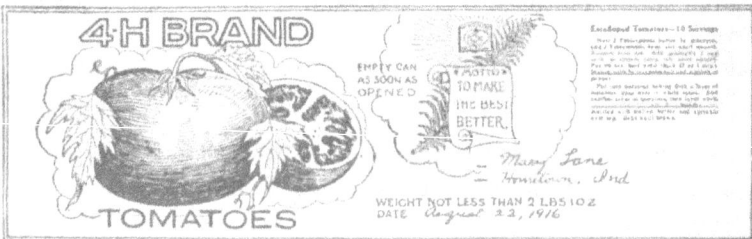

Reproduction of 4-H Brand Label Used by U. S. Department of Agriculture

46

SOLDER-SEALED TIN CANS

The hand-sealer is so simple and the open-top can so satisfactory that we prefer to use this method when possible.

But no one need hesitate to undertake canning in tin because soldering seems a complicated process. Soldering is simple and easy and instructs in a needed art. It is a distinct achievement for any boy or girl to be able to solder not only tin cans used in canning, but the leak in the water pail, the boiler, the dishpan, the basin, or other utensils used about the home. Soldering should be taught for its many other uses even if we are not using the solder-sealed can.

For solder-sealed cans, we shall need several things which we do not use when canning in glass, or with seam-sealed cans. In order to be sure that we overlook nothing here is a list which we can check off as we provide the material:

Tin Cans and Solder-Hemmed Caps— Solder-hemmed caps are not included with the cans but should be ordered extra. There is a binding of solder around the cap rim. The hot iron melts this and forms the seal.

Wire Solder—Our can caps are solder-hemmed but we need a small amount of wire solder to tip the vent. **A Capping Steel** to seal the cap. **A Tipping Copper** to melt the solder for closing the vent. **Sal Ammoniac** for cleaning the steels. **Soldering Flux—**The prepared paste, or, Powdered Rosin, or, Muriatic Acid and Zinc.

The Solder-Hemmed Cap

Prepared soldering flux can be secured at any tin shop or plumber's. Powdered rosin may be used, or, it is well to have on hand a small amount of muriatic acid and a few zinc chips such as can be picked up where the workmen have been laying a metal roof, or soldering water pipes. Then, in an emergency, we can make our own flux.

Making Flux—The acid-zinc mixture is prepared as follows: Take a small quantity — 10c worth — of muriatic acid. Cut the zinc into small pieces (not more than one-half inch in diameter). Drop into the acid all the zinc it will dissolve. Let it stand until it quits sizzling, then pour off the liquid, dilute it one-half, and bottle for use in roughening the tin so the solder will hold.

TINNING THE SOLDERING TOOLS

Lay the Solder on the Sal Ammoniac in a Circle

It is very difficult to solder cans smoothly and effectively unless our tools are clean and bright. If the steel itself is not covered with solder, the solder sticks and runs over the steel, and does not make a good job of the can.

Untinned steels should be tinned in advance, ready for use.

Tinning the Capper—Break several strips of solder into short pieces and lay them in a circle on top of a large lump of sal ammoniac, or, if powdered sal ammoniac is used, put 10c or 15c worth into an old bowl, or a tin can cut down to about half its original height, and lay the solder on this.

47

Have ready a weak solution of sal ammoniac dissolved in water.

File and rub off the dirt and rough places on the steel until it is smooth. If no file is convenient, use a soft brick.

Heat the Steel Very Hot, Dip It Into the Sal Ammoniac solution to clean it of smoke, or any particles which may have adhered to it. **Set the Steel on the Solder** and sal ammoniac, pressing down and **Turning It Back and Forth** until all the surfaces are bright. The hot steel melts the solder, the sal ammoniac cleans the steel and makes the solder flow smoothly over it.

Continue pressing and turning until the lower edge of the steel is covered with solder.

Tinning the Tipper—The process of tinning the tipper is much the same as tinning the capper.

The solder may be laid on a lump of sal ammoniac, or a little powdered sal ammoniac may be placed on a bit of cloth, and bits of solder mixed with it.

The tipping tool, which is usually made of copper, should be rubbed or filed, to clean off all dirt and rough places. Heat it very hot, dip it in the sal ammoniac solution to cleanse it, then rub back and forth on the solder and sal ammoniac, turning it over and over until the entire surface is covered with the solder.

Turn the Capping Steel from Right to Left Until the Lower Edge is Covered With the Solder

It is not necessary to tin the tools every time they are used, but they should be tinned often enough to keep them bright.

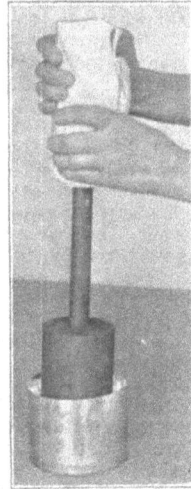

GENERAL INSTRUCTIONS FOR SEALING

Set Cans Level — If cans are set slanting, the solder will run to one side and the contents may touch the cover at some point and so render sealing difficult.

Wipe Cans and Caps Clean and Dry — Any foreign substance will interfere with perfect sealing.

Apply Flux to Cap: Flux Paste — Prepared, non-acid flux which can be purchased from any hardware dealer or plumber is a very convenient form. This should be applied by hand to the under edge of the caps before they are placed on the cans.

Flux Paste will not flow as a liquid flux will and must be applied at the point where it is wanted. When the cap is put in place, the flux is where the solder meets the tin.

Liquid Flux — If we use liquid flux, put the cap in place, dip a small brush in the acid, and wipe quickly around the edge of the cap. Enough of the liquid will penetrate beneath the cap.

Applying Flux Paste to the Cap

Be Sure Steels Are Well Tinned — Do not try to work without a well-tinned capper and tipper, or the solder will spread over the iron, instead of staying where it is wanted, at the joint of the can.

Have Steels Piping Hot — Experiment with a bit of solder. Notice how readily it runs with a hot steel, and how it clogs and lumps and is unmanageable if the steel is cold; then you will understand the necessity of hot irons to work with.

Try to Make a Neat Joint—If the steel is hot and we work quickly, it is as easy to make a smooth, neat job as it is to make a poor one.

Capping Tin Cans

A neat-looking can sells more readily than a botched one. It looks businesslike, and it does not suggest trouble with sealing, nor spoiled goods—either of which suggestions reduces the price.

Do Not Try to Cap When Vent is Closed —Because of the large surface to be closed at once when capping, a small opening is left in the center of the cap to provide an outlet for the steam. Then when the cap is sealed, some solder and a hot iron is applied directly to close the opening the vent.

If after you have closed the vent you find a leak in the cap seal, punch a hole in the cap, make the cap seal perfect, then close the opening you made in the cover.

In tipping, or sealing the vent, the solder is in direct contact with the opening and the hot steel; in capping, the body of the cover is between the iron and the solder which rests on the can; that is why it is easier to seal the vent than it is to mend an imperfection in the cap solder.

Test All Seals After Sterilizing—Stand cans cap-side down to cool. If you find a leak, punch a hole in the cap, solder the leak, then seal the vent made, and return the can to the cooker for five minutes.

CAPPING TIN CANS

Start heating the soldering tools in time so that they will get very hot by the time you are ready to use them.

The self-heating capper which has a small gasoline burner attached, is very satisfactory. It is light in weight, always hot, and prevents delays.

If we are using the ordinary capping tool which must be heated over the fire, we place it over the gas burner, on the bed of coals, or in the plumber's firepot, in plenty of time to heat it thoroughly through and through. The center rod is removed when we are heating the steel.

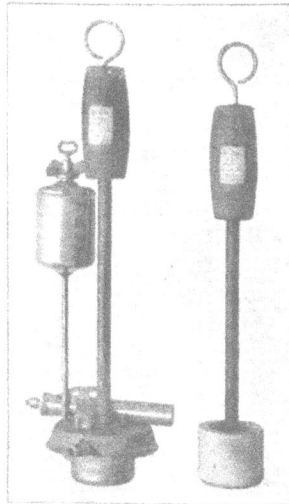

Two Styles of Cappers. The One on the Left is Self-Heating

Assume that our cans are filled, the flux has been applied, the cap is in place, and our steel is hot:

Lift the Hot Steel with the Right Hand—Remember, the capping tool is heavy — it must be an effective tool for sealing — and now it is hot. Handle it carefully, not to strike anyone or anything, and do not drop it.

Put Rod in Place with Left Hand —The rod serves as a guide in handling the steel and may prevent accidents.

Dip the Steel in the Sal Ammoniac Solution —This is to clean the steel of smoke and particles which may have attached to it while heating, and so make the solder flow more smoothly.

Set Tip of Rod Over Vent — The lower end of the guide rod in the usual form of capping steel, has an inverted v-shaped (Λ) end, the two points of which are set on opposite sides of the vent hole.

Lower Steel to Can —That is, lower the capping steel itself until it rests on the solder — be sure it touches the cap rim at all points.

Give It Two or Three Quick Turns back and forth, then,

Hold the Capper a Second to Allow the Solder to Set

Raise Steel an Inch and Hold an Instant to Let Solder Set

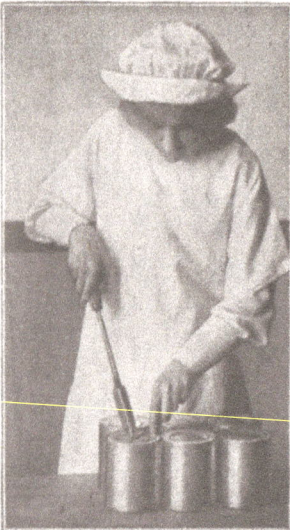

Press the Solder Against the Point of the Tipper

Cover Vent, Invert Can, and Watch for Leaks.

TIPPING TIN CANS

When canning in tin, we seal the cans at once. The tin will bulge out in cooking, but is strong enough to withstand the pressure, and when the contents cool, the can will come back into shape.

See That Cap Seal Is Perfect — It is impossible to repair leaks in the cap solder after the vent is closed. For this reason it is important to know that our cap seal is perfect before we undertake to close the vent.

As in the case of the capping tool, we have our tipping copper very hot. Dip the tipper in the sal ammoniac solution to clean it, hold the solder with the tip in the vent, press the steel to the solder, remove the solder quickly, and, if necessary, smooth the drop on the can with the tipping copper to make a smooth seal.

Work Quickly — Quick work is required to produce a good, neat seal.

CANNING REMINDERS

Begin with one product only.

Experiment with a small quantity.

Read carefully the instructions for canning tomatoes found on pages 26 to 32.

Do not try to follow two sets of instructions. Follow one faithfully.

Do not can a large pack without trying a jar or two to see that the seasoning and sterilizing have been properly done.

It is important that the instructor when giving a canning demonstration should have the necessary material, and that the equipment is in working order, products on hand, outfit in good repair, jars clean, and everything ready for rapid work and accurate results.

If you grow fruits and vegetables for canning, grow the varieties which when canned are of good color, flavor, and texture. Color, flavor, and texture affect palatability and price.

Remember, that you can successfully can any product by the cold pack method — not only fruits and vegetables — but meats as well.

In a week's work with a canner, it is possible to can a can of fruit, a can of vegetables, and a can of greens for each day of the year. Three hundred cans of tomatoes (or other quickly prepared, quick-cooking product) per day is not an uncommon task for one girl, even a small girl, with a canning outfit.

Always Blanch and Cold Dip All Vegetables—All vegetables should be blanched. Any fruit or vegetable that is blanched should be immediately plunged into the cold dip. A product may be blanched in boiling water or in live steam.

In addition to its influence in the keeping of vegetables, blanching shrinks the product by driving the gases from the tissues. This space absorbs liquid when cooking, thus plumping the product and making it crisp and of better appearance. Try canning some snap beans and some apples blanched, and some unblanched, and see for yourself the advantage of the blanch for the appearance alone.

A Collection of Handy Utensils for Canning Which Any Woman Can Have

51

Cleanliness — Absolute cleanliness is necessary, for health, and for palatability and in order to prevent danger of spoiling. A dirty pack will contain a large number of bacteria. The larger the number of bacteria, the more likely the product is to spoil.

Canning Outfits — While a regular canning outfit is an advantage, especially if one is going to can to sell, it is not necessary to buy either outfit or cans. An outfit will cost from $15 to $20.

Any clean jars or cans which are on hand may be used, and the pack cooked in a pail, kettle, boiler, or any sort of clean vessel deep enough so that the cans may be covered with water.

Canning in Glass — If the covers to glass jars are screwed too tight, the rubbers will be forced out of place; if too loose, the water may exhaust. A rubber which bulges out may be too large. In that case substitute a new rubber and cook for five minutes.

If the rubber bulges because the cover is screwed down too close, simply loosen the top, slip the rubber back in place, and tighten.

In using glass jars use ordinary common sense in handling them to prevent breakage.

By using hot jars and hot syrup or hot water for filling, jars may be set directly into boiling water. The hot jars also hasten the cooking.

Cooking — Different seasons produce different products. In extremely dry seasons many of the bacteria are transformed into spores, which are more difficult to kill than the ordinary bacteria.

Some products need to be cooked quickly, and so are best canned at a high temperature; the delicate flavor and texture of some is spoiled by intense heat, and such products are best when given a longer period of sterilization at a lower temperature.

For instance, peas should never be cooked at a steam pressure above 10 lbs., although corn may be canned at 15 lbs. pressure.

Over-processing is apt to give some products, such as sweet potatoes, pumpkin, and squash, a scorched taste and appearance.

Excessive shrinkage, an abundance of liquid in a can which was properly packed, or a feathery appearance, indicate over-cooking.

Under-ripe and over-ripe products and products canned without sugar need longer processing.

In general, the regular instructions will produce an article that will keep and be salable, but remember that color, flavor, and texture affect palatability and price. If you wish to get fancy prices you must study your product, use judgment, and produce canned goods which, to the main essential that they keep, add the qualities of delicate flavor, attractive color, and firm texture.

Time-table — Boiling temperature varies at different heights, and in high altitudes the time for cooking in a *hot water bath outfit* must be increased as follows :

> 500 to 1500 feet, use time-table as given
> 1500 to 3000 feet, add 10 per cent
> 3000 to 4000 feet, add 20 per cent
> 4000 to 7000 feet, add 40 per cent

Labeling — It is important to label all goods. For tin cans, use the regular commercial label which fits around the entire can. The labels may be laid face down on the table, over-lapping so that the edges are exposed. With a large brush apply paste to the entire lot at once, simply pasting across the ends. Pick up a label, lay it around the can, overlap the edges and press them together so that the paste holds it in place.

A labeling contest is one of the amusements sometimes planned by canning clubs.

Storing — When the product is taken from the cooker do not set tin cans in the hot sun, or a hot room, nor pack them together too close or they will retain the heat and overcook. Do not store in a damp place.

To retain color and texture do not expose canned products to the light. If canned in glass, wrap in paper.

Recipes — It has not been thought advisable to print a number of recipes. There are so many products to can that it would be impossible to include them all without making this booklet so bulky that it would not be valuable as a hand-book.

Most companies which manufacture canning outfits furnish recipe books and the leaflets distributed by the canning leaders in the Office of Extension Work, U. S. Department of Agriculture, Washington, D. C., may be secured free of charge.

SOME SPECIALS

Use Lacquered Cans or Glass Jars for Very Acid Products — Cherries, blackberries, raspberries, all red fruits, gooseberries, pumpkin, beets, squash, sweet potatoes — these products lose color if canned in tin.

Rhubarb Should Always be Canned in Glass — It contains a very strong acid which will affect even the lacquered tin.

Acid Products — Tomatoes, rhubarb, gooseberries, and other fruits and vegetables with a high percentage of acid keep most easily. Such fruits and vegetables shrink most in canning.

Products Which Deteriorate — Apples and blackberries lose quality with age and should not be kept over from year to year. It is best to market them after canning.

Fruits Which Mould or Work may sometimes be saved if treated at once. Immediately they show signs of spoiling *loosen the covers* and cook *in the jar* for 10 minutes or longer as indicated.

Increasing cloudiness of liquid or fresh bubbles indicate spoilage.

Do not sell fruit which has been reheated to check spoilage.

Do not try to save vegetables which show signs of spoilage.

Protein Foods — Protein is a favorable medium for the growth of bacteria. Corn, beans, peas, pumpkin, squash, and sweet potatoes — all vegetables with a large protein content — require a high degree of heat or a longer period of sterilization. These products expand in cooking. Do not fill the cans too full.

Wilted Vegetables should stand uncut in cold water until they are crisp.

Apples — Blanching greatly improves the texture and appearance of apples.

Apples and some other fruits have a tendency to turn brown when allowed to stand after they are cut. To prevent them discoloring, the pieces may be dropped into mild salt water, as they are pared and sliced. Let them stand for five minutes, then wash in clear water and pack.

Use a thin syrup.

Summer apples are not firm enough to keep well when canned. They cook up and lose flavor. They may, however, be canned to be used in a short time.

Windfall apples may be pared, cored, and sliced, using water, and only a small quantity of that, instead of syrup, and canned for pies.

The No. 10, or gallon, tin can is usually preferred for apples.

It is suggested that housewives who can in glass will have used a part of their canned goods and have a number of empty cans on hand soon after the holiday season. At this time the winter store of apples often begins to decay. Apples which will not keep uncanned may be canned in the empties and kept for late winter and early summer use.

Beets — To retain the color of beets leave three or four inches of the stem and all the root on while blanching. Blanch in steam instead of water. After blanching, the skin may be scraped off.

Corn — In canning corn on the cob select Golden Bantam, Country Gentleman, or some other small-cob corn, to save space.

If the corn is too ripe it becomes dry and discolored while processing; if it is under-ripe it is tasteless and lacks food value.

Be careful not to use too much salt in corn, as it seems to cause it to develop a "sour" taste. A small spoon of a mixture of two-thirds sugar and one-third salt is considered a good proportion.

Very hard water sometimes causes corn to turn yellow and may also spoil the flavor. Immature corn will sour more readily than corn which is at just the right stage.

It is best to can corn within a half hour after gathering, but if pulled with husks and a considerable piece of the shank left on, it will keep fresh for some time. Corn which has been gathered for some time is more liable to spoil.

Corn on the Cob — Husk, silk, and trim, cutting out any poor kernels. Cut off the tips of the ears if necessary to get them in the can. Do not leave any broken kernels, as they will give a milky appearance to the water in which the corn is canned.

Blanch as per time-table, plunge into cold dip, and pack quickly, alternating tips and butts — first ear, tip end down; second ear, butt end down — and so on, so that they fit closely in the can and no space is wasted.

A quart Mason jar will hold four ears of Golden Bantam. Gallon (No. 10) tin cans are best for canning corn on the cob. They hold from seven to twelve ears each.

Salt and add from one to two inches of water. Corn looks better if the can is filled with water, but it tastes better if only a small amount is used.

When using corn canned on the cob, take the ears out of the liquid and put them in a steamer and steam until heated through, then lay in a medium hot oven for a few minutes to dry out before serving. If the ears are heated in water the corn is apt to taste watery.

Canning Corn Cut Off — Blanch on the cob as per instructions for canning corn on the cob. Cold dip, cut off (drawing the knife from the tip towards the base of the cob), pack, salt, and add a small amount of water.

Unlike most other products, corn swells in cooking so the cans should not be packed too full. Leave one-half to three-quarters of an inch of space at the top of the can.

Cauliflower, Cabbage, and Sauerkraut should be soaked in cold salt water 8 to 6 hours.

Greens — Blanch all greens in steam. Blanch or cook twenty minutes to reduce bulk. Pack close. Can in glass or lacquered tin.

Rhubarb — Blanch rhubarb before peeling.

Never can rhubarb in tin cans. Rhubarb contains an especially strong acid which will eat even the enamel-lined tin cans.

Squash and Pumpkin should be cut into sections, blanched 10 minutes in the shell, cold-dipped, then scraped out of the shell, packed and cooked as per time-table.

Can in glass or lacquered tin.

FRUIT JUICES

Fruit juices furnish a healthful and delicious drink and are readily canned at home. Each home supply room should have, not a few quarts but an abundant supply, of canned fruit juices which, in addition to supplying flavoring for puddings, gelatins, etc., may be used freely as a beverage.

Grapes, raspberries, and other small fruits may be crushed in a fruit press, or put in a cloth sack, heated for 30 minutes, or until the juice runs freely, and allowed to drip.

Strain through two thicknesses of cotton flannel, to remove the sediment, sweeten slightly, bottle, close by filling the neck of the bottle with a thick pad of sterilized cotton, heat to 160°, or until air bubbles begin to form on the bottom of the cooker, and keep at this temperature 1½ to 2 hours; or, heat to 200°, or until the bubbles begin to rise to the top of the water, and hold at this temperature for 30 minutes. Cork without removing the cotton. If canned in jars, close the jar partly, the same as when canning fruits and vegetables, and seal tight after cooking.

Fruit juices should never be heated above 200°, as a higher temperature injures the flavor.

A very good quality of grape juice may be made by selecting perfectly sound, whole grapes, picking them from the stems, washing them through several waters, then canning them as follows:

Place one pint of grapes in a 2-qt. jar, add ½ cup sugar, fill the jar with boiling water, and seal tight at once. It is not necessary to cook this.

Apple Cider may be bottled, heated to 180°, and held at this temperature for 45 minutes.

A small portion of grape, currant, or blackberry juice added to canned apple cider when it is served restores its pungency. Pouring it back and forth from one pitcher to another just before serving, so it can absorb air to take the place of that driven out by heating, also brightens its flavor.

SOUPS AND MEATS

Soup stocks, purees, consommes, and vegetable or meat soups are readily canned, and are palatable and economical.

Meats may be canned instead of corning or smoking, or corned meat may be canned. Chicken Fries canned in the late fall preserve the meat at the most delicious stage and we avoid the expense of feeding throughout the winter the chickens intended for the family meat supply. Game and fish may be canned to serve as a delicacy at a time of the year when it may be difficult or even impossible for most of us to secure them otherwise.

Be sure that meats for canning are in perfect condition.

Meats should be cooled quickly, the bone, gristle, and fat removed, then cut into convenient pieces. Sear and pack at once. Fill the jar with hot "pot liquor," or boiling water, season as desired, cover, and cook as per time-table.

Tough meats, old fowls, and other meats which require long cooking to make them tender, may be boiled a half hour or longer before packing.

Fish should be soaked in brine a half hour before packing.

Too high temperature injures the flavor, destroys the texture, and shrinks meat. For this reason many people prefer to can meat in a hot water bath instead of a steam outfit.

Write direct to the Office of Extension Work, U. S. Department of Agriculture, Washington, D. C., for detailed recipes.

SYRUPS FOR CANNING

Syrups are used in canning most fruits.

Time is saved by using hot water in making syrups, as the sugar dissolves more readily.

In making the syrups given here, stir the sugar into the water, and let the syrup come to a bubbly boil. Boiling is not necessary; it only makes a thicker syrup. Do not stir after the sugar is thoroughly dissolved.

Thin Syrup—1:2, that is, one cup of sugar to two cups of water. Use for peaches, apples, and fruits that are not delicate in texture and color.

Medium Syrups—1:1, that is, equal parts of sugar and water. Use for blackberries, currants, blueberries, huckleberries, black raspberries, cherries, plums, etc.

A 3:2 syrup, three cups sugar to two cups water, makes a *medium thick* syrup. Use this for strawberries, red raspberries, and especially sour fruits, such as gooseberries and cranberries. The thicker syrup helps to preserve color and texture.

Thick Syrup—2:1, two cups of sugar to one cup of water, for sun preserves and jams.

The thickness of the syrup will vary with the variety of fruit and with the taste of the individual.

Most fruit must be sweetened before it is used, and if sweetened when it is canned, it has a better flavor than if sweetened when served.

JELLIES AND PRESERVES

The best jellies are made in the proportion of three-quarters cup sugar to one cup fruit juice. More sugar makes more jelly but it does not stand up as well; less sugar makes a tough jelly.

Pectin is the principle which makes fruit juice jell. It is found in most fruits and some vegetables. Apples, the whites of the citrus fruits, and carrots contain an abundance of pectin, that is why we add apple juice to some fruit juices which do not have sufficient pectin to jell alone.

It is not practical for the housewife to make pectin, but commercial pectin is now for sale and a small amount of it added to the juice of fruits which do not jell readily makes jelly-making certain.

Where pectin is used we depend upon the fruit to furnish coloring and flavor; the amount of jelly secured depends upon the amount of sugar used; that is, so long as there is enough pectin to use the sugar. Write the Office of Extension Work, U. S. Department of Agriculture, Washington, D. C., for recipes for making jelly with pectin.

Sun Preserves—Strawberries, raspberries, ripe gooseberries, cherries, etc., make good sun preserves. Peaches sliced or cubed are also good.

Select the fruit, sprinkle lightly with powdered sugar, cover with the thick syrup and set in the sun. Protect from insects, but do not cover close with glass, as this retains the moisture and prevents the proper cooking of the fruit.

A south wall for a background helps concentrate the heat.

Preserves from Dried Fruits — Dried fruits, such as apricots, peaches, etc., make very excellent preserves. They have a distinct flavor and are richer than when fresh fruits are used. Soak the fruit over night in a small quantity of water, then proceed as with fresh fruit.

Jellies, Jams, Preserves, and Fruit Butters do not need to be sealed, as there is enough sugar added to preserve them. They may be canned in open glasses or jars, and the top covered with melted paraffin. If desired, a small piece of paraffin may be placed in the bottom of the jelly glass when the jelly is poured in. The paraffin will float and will be melted by the heat of the jelly and form a perfect air-tight seal. The jars or glasses should be covered when cold with tin caps or with paper, so that dirt and dust will not collect on the food. A small rubber band may be snapped around the neck of the jar or glass to hold the paper in place.

CANNING TO SELL

If canning to sell, write to the Pure Food Commission, or Health Department, of your own state, and to the Bureau of Chemistry, U. S. Department of Agriculture, Washington, D. C., for copies of the Pure Food Laws and Regulations concerning canned goods to be sold.

Put your name—a trade name if desired—and address on each can so the buyer will learn to know your brand. Make the food so good that the customer will re-order. Canvass your trade in advance so that you will have a market for your products.

Cater to high-priced trade. Sell only first-class canned goods. See that the container and the label are attractive, then ask a fair price.

Hospitals, Colleges, Boarding Houses, Hotels, Railroad Diners, the Neighbors, and the Home Grocer are all possible customers.

EXHIBITS

In preparing canned goods for exhibits, see that the cans are all of one size and make. This insures a uniformity that makes a better looking exhibit. Tops should be new and bright and the cans scrupulously clean and polished. A dark green crepe paper for a background and some ferns and flowers set among the jars add to the appearance.

HOME CANNING CLUBS

More than 500 Club Leaders and Home Demonstrators, working under the direction of the U. S. Department of Agriculture, are helping to spread the story of how simple and easy it is to do one-period, cold-pack canning at home.

There is no reason why there should not be a Canning Club in every district of every State of the Union, affording the farm girl an opportunity to earn money, to develop her business ability, and to meet in the social gatherings which grow out of Canning Clubs.

It has not been found advisable to organize a county in the club work unless the local authorities co-operate by appropriating a part of the money necessary to pay the salary of a County Agent.

The State Colleges of Agriculture co-operating with the U. S. Department of Agriculture are now paying part of the salary of a local or district leader in some communities where the organization is satisfactory.

The first thing to do is to work up enough local sentiment so that local funds are available, then present the matter to the Director of Extension in your state, or write the State Club Leader in care of the Extension Director, State College of Agriculture, your state, and learn what steps are necessary.

HISTORY OF THE HOME CANNING CLUBS

The first Girls' Tomato and Canning Club was organized at Aiken, South Carolina, in 1910, by Miss Marie Cromer, a teacher in the rural schools. It was intended to give girls in country districts an opportunity similar to that which the Corn Club offers to boys. Miss Cromer, who is now Mrs. Seigler, was assisted in planning the details of the work by County Superintendent Cecil H. Seigler.

Dr. Seaman A. Knapp, the great agricultural educator, was at that time Special Agent Farmers' Co-operative Demonstration Work, with the U. S. Department of Agriculture. He saw the value of this work both in saving food products now wasted and as a training school for girls, and promptly sent the Club a canning outfit, cans, and labels. Secretary of Agriculture James Wilson added a check for $100 and with this financial assistance forty-six girls began Home Canning.

Mrs. Marie Cromer Seigler, who organized the first Canning Club

The first season they canned by the Cold Pack method more than 6,000 cans of tomatoes and many gallons of catsup and other products. Within a year 325 girls were enrolled and the work had spread to other states. In 1912, its value had become so apparent that it was decided to extend it through all the states, and now there are more than 500 demonstrators and several hundred thousand members.

Since the adoption of the one-period process, simplifying the work and shortening the time required, Cold Pack Canning has come into more general use, and it is estimated that more than 500,000,000 jars of canned goods were packed last year by home workers.

The work is not confined exclusively to girls, but boys, too, are often included in the club, and within the past year, the Mother-Daughter Clubs have been organized. These give the women of the community an opportunity to train in this work.

The First Tomato Club, Aiken, S. C., 1910

58

SAVE EVERYTHING—WASTE NOTHING
Can, Dry, Store, Pickle or Preserve

No product should be wasted. There is no reason why any of it should be lost.

There isn't a thing grown in the garden or orchard that we cannot save in some way. We can can it, or pickle it, or dry it, or bury it in the ground or in sand or sawdust in the cellar, or simply put it in the cellar.

Waste, bad enough at any time, is criminal under present conditions, for we will need all our products for the soldiers at the front and those of us at home.

Waste is bad management; saving is profitable.

If we cannot get jars or cans enough for canning, we can save fruits and vegetables in other ways. Some things are better preserved without being canned.

Many products can be pickled; others can be dried; others can be stored in cellars or buried in the ground.

Peas and beans that get too ripe for canning should be dried and hung up in sacks in the cellar. Even though there may be but a quart or two, they will help feed us and will be wholesome.

Turnips, beets, carrots, radishes and parsnips may be kept in moist sawdust or sand in the cellar or in a pit outdoors.

Cabbages may be kept in an outdoor trench or in barely moist sand in boxes or barrels in the cellar.

Sweet potatoes can be stored. An easy and effective way of storing them is in use in the South. A bulletin describing the method will be gladly sent free to anyone addressing this department.

Cucumbers, beets, cauliflower, snap beans, green tomatoes, small white onions and melon rinds may be pickled.

Pumpkins and squashes can be kept for a while in a warm, dry part of the cellar or may be cut up and canned or dried.

Potatoes can be kept in a pit out of doors or stored in a dry bin in the cellar, where it is not warm enough to cause them to sprout or cold enough to freeze them.

Rhubarb should be canned, and, after the ground has been frozen in the fall, a few plants may be dug up and transplanted in the cellar, where they will grow all winter if the temperature is not too low.

Blackberries, raspberries, blueberries, currants and cherries may be dried by spreading them on a piece of canvas fastened to lath and suspended above the kitchen stove, a piece of mosquito netting being used to protect them from flies, or they may be made into jelly or jam.

Apples may be stored in barrels or bins, or may be dried, canned or made into sweet pickles. Even the windfalls should not be allowed to go to waste. They should be cut up and canned, made into apple butter or cider, or preserved.

Peaches that are not fit to can may be saved in much the same way as apples.

These are some of the ways in which we may save all the products of the orchard and garden. There is a way to save everything and none of these products should be wasted.

Can, Dry, Store, Pickle, Preserve, or Bury. Save everything. Let nothing go to waste. We will need it. Our country will need it.

Every family should have a copy of Farmers' Bulletin, No. 841, "Drying Fruits and Vegetables in the Home." Send to Division of Publications, U. S. Department of Agriculture, Washington, D. C., for a free copy.

59

TIME-TABLE for COLD PACK CANNING

For Scaldi ngor Blanching, and Sterilizi ngin Cold Pack Canning

Use the Time Given Under the Type of Outfit You Are Using.
See note under "Time-Table," Page 52.

PRODUCTS	*Syrups	Scald or Blanch	Hot Water Bath Outfits at 212°	Water-Seal Outfits 214°	Steam Pressure 5 Lbs.	Pressure Cooker, 10 to 20 Lbs.
Fruits						
Apricots	1 S: 1 W	1 to 2 Min.	16 Min.	12 Min.	10 Min.	5 Min.
Blackberries	1 S: 1 W	No	16 "	12 "	10 "	5 "
Blueberries	1 S: 1 W	No	16 "	12 "	10 "	5 "
Cherries	1 S: 2 W	No	16 "	12 "	10 "	5 "
Cranberries	3 S: 2 W	No	16 "	12 "	10 "	5 "
Currants	1 S: 1 W	No	16 "	12 "	10 "	5 "
Dewberries	1 S: 1 W	No	16 "	12 "	10 "	5 "
Gooseberries	3 S: 2 W	No	16 "	12 "	10 "	5 "
Grapes	1 S: 2 W	No	16 "	12 "	10 "	5 "
Peaches	1 S: 2 W	1 to 2 Min.	16 "	12 "	10 "	5 "
Plums	1 S: 1 W	No	16 "	12 "	10 "	5 "
Raspberries	1 S: 1 W	No	16 "	12 "	10 "	5 "
Rhubarb	1 S: 1 W	1 to 2 Min.	16 "	12 "	10 "	5 "
Strawberries	2 S: 1 W	No	16 "	12 "	10 "	5 "
Citrus Fruits	....	1½ Min.	12 "	8 "	6 "	4 "
Apples	1 S: 2 W	1½ "	20 "	12 "	8 "	6 "
Pears	1 S: 2 W	1½ "	20 "	12 "	8 "	6 "
Pineapple	1 S: 2 W	10 "	30 "	25 "	25 "	18 "
Quince	......	6 "	40 "	30 "	25 "	20 "
Figs	2 S: 1 W	15 to 20"	40 "	30 "	25 "	20 "
Some Specials						
Tomatoes		1 to 3 Min.	22 "	18 "	15 "	10 "
Tomatoes and Corn		T. 2, C. 8	1½ Hrs.	1¼ Hrs	1 Hr.	45 "
Egg Plant		3 Min.	1 "	45 Min.	45 Min.	30 "
Pumpkin		5 "	1½ "	50 "	40 "	35 "
Squash		5 "	1½ "	50 "	40 "	35 "
Corn (on cob or cut off)		5 to 8 "	3 "	1½ Hrs.	1 Hr.	45 "
Hominy		5 "	1½ to 2 "	1½ "	1 "	40 "
Greens, Roots, Tubers and Other Vegetables						
Dandelions		10 to 15 Min.	2 "	1 "	50 Min.	25 "
Spinach		10 to 15 "	2 "	1 "	50 "	25 "
Greens, all other kinds		10 to 15 "	2 "	1¼ "	1 Hr.	35 "
Asparagus		2 to 4 "	1½ "	1 "	50 Min.	25 "
Beans (lima or string)		5 "	1½ to 2 "	1½ "	1 Hr.	40 "
Okra		5 "	1½ to 2 "	1½ "	1 "	40 "
Peas		5 "	1½ to 2 "	1½ "	1 "	40 "
Brussels Sprouts		4 to 10 "	1½ "	1 "	50 Min.	25 "
Cabbage or Sauerkraut		6 to 15 "	1½ "	1¼ "	1 Hr.	35 "
Cauliflower		3 to 6 "	1½ "	1 "	50 Min.	25 "
Beets		6 "	1½ "	1¼ "	1 Hr.	35 "
Carrots		6 "	1½ "	1¼ "	1 "	35 "
Sweet Potatoes		6 "	1½ "	1¼ "	1 "	35 "
Parsnips, Turnips, etc.		6 "	1½ "	1¼ "	1 "	35 "
Meats and Soups						
Beef and Pork		See Page 55 for Tough Meats	4 "	4 "	3½ "	1½ Hrs.
Poultry			4 "	3½ "	3 "	1 "
Fish and Shell Foods		3 to 8 Min.	3 "	2 "	1½ "	1 "
Soup Mixtures		3 to 8 "	1½ "	1¼ "	1 "	30 Min.

*"S." indicates 1 part sugar.
"W." indicates 1 part water.

CANNING LITERATURE

Canning Literature Furnished by the United States Department of Agriculture, Washington, D. C.

Farmers' Bulletin 839 — By O. H. Benson. (This is the Most Complete Publication by the Government on Canning).

Additional Recipes, Tested and Determined for Use in the Boys' and Girls' Home Canning Club Work — By Geo. E. Farrell.

Tinning, Capping and Soldering: Repair Work for Farm Home — By O. H. Benson.

Suggestions for Organization of Mother-Daughter Home Canning Clubs.

Suggestions No. 4 to Club Leaders and Demonstrators in Home Canning Club Projects— By O. H. Benson.

Home Canning Club Instruction — Canning of Soups — By Geo. E. Farrell.

Canning Windfall and Cull Apples and Use of By-Products — By O. H. Benson.

Canning Tomatoes at Home and in Club Work — By J. F. Breazeale, Washington, D. C.

Home Canning Instructions — By O. H. Benson, Washington, D. C.

Home Canning Club Instructions to Save Fruit and Vegetable Waste.

Home Canning Directions to Help in Overcoming Common Difficulties.

Canning Clubs in New York State, Part 2.—Principles and Methods of Caning — Cornell Reading Course Vol. 3 No. 69. Flora Rose—O. H. Benson.

Formula for Removing Skins from Peaches, Plums, Pears, Etc.

Home Drying Manual for Vegetables and Fruits—Pub., by The National Emergency Food Com., Washington, D. C.

61

THE FOLLOWING IS FOR HISTORICAL PURPOSES ONLY
AND DOES NOT REFLECT CURRENT
SCIENTIFIC KNOWLEDGE, POLICIES,
PRACTICES, METHODS OR USES.

AN

ACRE HOME

FOR

EVERY

AMERICAN

Copyrighted 1919

PUBLISHED BY

INTERNATIONAL HARVESTER COMPANY
(Incorporated)

AGRICULTURAL EXTENSION DEPARTMENT
P. G. HOLDEN, DIRECTOR

HARVESTER BLDG., CHICAGO

27 A—4-26-19.

PATRIOTISM

Breathes there a man with soul so dead,
Who never to himself hath said:
"This is my own, my native land!"
Whose heart hath ne'er within him burned,
As home his footsteps he hath turned
From wandering on a foreign strand?

If such there breathe, go, mark him well;
For him no minstrel raptures swell;
High though his titles, proud his name,
Boundless his wealth as wish can claim;
Despite those titles, power and pelf,
The wretch, concentred all in self,
Living, shall forfeit fair renown.
And, doubly dying, shall go down
To the vile dust from whence he sprung,
Unwept, unhonored, and unsung.

—SIR WALTER SCOTT.

THE HOME OUR BIG PROBLEM

Ownership Means Real Citizenship—
Lack of Homes Means Anarchy

There must be a home for every American and an acre of ground with every home.

Unless we want to go through the red fire of anarchy that has destroyed Russia, we must adopt some plan that will result in making a large majority of our people home owners.

The greatest problem facing America today is that of a higher standard of citizenship, and the greatest elements in the development of that higher type are, first, the owning of some property, especially a home; second, the forming of the habit of industry in our future men and women. Given these two things, America need not worry about outlawry or anarchy.

The only real home is that which is rooted in the soil. If we can help a man to own even a modest cottage and an acre of ground, we are doing a real service to the state.

If every family is to have a home, we must make it possible for the man with a small wage and a family of three or four children to buy a home and pay for it.

Garden Will Help Pay for Home

If we help such a man to get hold of some land where he can have a garden, some fruit and some chickens—where the children can help produce from one-third to one-half the family's food, and perhaps have some surplus to sell, we will be making him and his family an asset to the community and the state.

And this can be done. If we are willing to help these people as our citizenship demands that we should help them, we can make it possible for them to buy and pay for their homes as easily as they can pay rent.

Suppose a workman with a family of five has a salary of $1,000 a year. He cannot rent rooms fit to live in for less than $15 a month or $180 a year. Food for the family, all of which he must buy will cost him 40 per cent of his wage, or $400 a year. His rent and food cost him $580.

Suppose we make it possible for him to buy a house and an acre, or even half an acre of ground for $2,000 and allow him seven years in which to pay for it, at an interest rate of 5 per cent.

His annual payments on the principal will be $286 and his maximum annual interest will be $100. His garden, his chickens,

5

his fruit, will reduce his expense for food one-half, or to $200. This will make a total expense of $586, practically th: same as if he rented. And this is not counting the amounts saved in doctor bills, nor is it taking into consideration the reduction each year in the interest he pays.

He should be allowed 10 years, 20 years or even 30 years to pay for his home, both principal and interest to be met in equal monthly installments. Such a plan would make it much cheaper for him to buy than to rent.

Personal Service Needed

This idea of a home for every family is the biggest idea any state or community can take up. It calls for service. We must do things that we are not paid for doing. Service is the price we pay for government and efficiency.

There are thousands of little towns in every state that should begin right now to think about this problem. If they follow a bad system of city building, it will be bad for all time.

Every organization, business and social, should get behind the movement. A great campaign should be held. There should be thousands of meetings covering every county in the state. A plan should be worked out that would meet with the endorsement of every organization. Service, not selfishness, should be the motive.

The Governor should issue a proclamation calling the mayors together. The mayors should call the people together to formulate a plan or policy looking into the future.

No Escape From Disease Germs Here

PRESERVE YOUR HOME   186

AN ACRE FOR EVERY FAMILY

Half a Million People Can Have Homes Within Radius of Seven Miles from Center of Town

How can every family have an acre of ground, you ask. There is not room; they will have to move out too far, you say.

Draw a circle seven miles in all directions from the center of your town. There will be 100,000 acres of land inside the circle. If your business district is two miles square it contains less than 3,000 acres. This leaves over 97,000 acres for homes—an acre apiece for 97,000 families. The average family consists of five people; this will provide homes for nearly a half million people. If each family has a half acre you can provide for a city of nearly a million inhabitants—all within seven miles of the business district.

The average tenement district has 200 families to an acre. Many have over 300. There have been cases of 100 families under one roof.

The economic thing is not to have a home with just a paved street in front and no ground—but a place big enough to produce something—to grow some of the necessities of life.

A House is Not a Home

You can buy or build a house but that is not a home. It is only a house. If the workman wants a real home, one in which he can live cheaper than he can rent, he must go far enough out to have some land so he can grow a garden and reduce the cost of living.

In nearly every family there is a grandfather or a grandmother, or children too young to work; perhaps there is some one in poor health. If there is a garden these will have something to do—something that will help support the family.

And then there is mother. Mother can work in the garden. It will keep her away from the factory; keep her the home-maker. When mother works in a factory the home life disappears.

A garden and some fruit will help feed the family. This will pay for father's transportation to and from work.

Such a home will be a real home. It will be a home the workman can buy and pay for. It will provide outdoor exercise, sunshine and pure air. It will keep the family in touch with growing things. It will improve their health; keep them vigorous and strong; make them good citizens.

7

IMPORTANCE OF HOME OWNERSHIP
It Means Better Health, Greater Efficiency, Less Vice and Crime

Home ownership means:

1. Better Health.
2. Greater Efficiency.
3. Less Vice and Crime.

Health—The expensive but unsanitary apartment house, the wretched hovel, the overcrowded tenement—these impair the physical and moral health of the tenant. They are unfit for living or homemaking. They are damaging to the community.

Every family has a right to sunlight, fresh air and pure water, yet in many cities, only those who can afford to pay for these things have them.

America has vast forests, mighty hills and mountains, broad plains with only here and there a cabin. We have unbroken prairies where the homes of men are far apart. Thousands of farms are so large that their owners cannot cultivate them as they should be cultivated; the fertile fields are lonely for the pressure of a human foot, and yet, in the great wilderness of tenement houses, where it is too damp and too dark for even grass to grow hundreds of human beings live upon a single acre.

Room for Every Family

In the suburbs of every great city there is room for every family to have a home and a garden. Even in Rhode Island, the most densely populated of all our states, there is at least two and one-half acres of land for every family.

For 6,000 years humanity has been building—and we have not learned how to make homes.

Children from hundreds of families are crowded together in dark, damp places. There is no sunlight; the air is foul; seeds of moral and physical infection are sown. Unless we plant something worth while and cultivate carefully, only weeds will grow.

Tuberculosis is frightfully common; typhoid is everywhere.

There is always sickness in the tenements. The only wonder is that more do not die.

"The most pitiful victim of modern city life is not the slum child who dies, but the slum child who lives. Every time a

8

baby dies, the nation loses a prospective citizen, but in every slum child who lives the nation has a probable consumptive; a possible criminal."

Disease, defectiveness, delinquency and dependency—these form the annual harvest of the tenements—these form the shame of the cities.

Humanity Our First Consideration

We expect human plants to thrive where vegetable plants wither and die. We grow our crops and herds in the pure air of the open fields and our boys and girls in the darkness' and dirt. And the most valuable thing on earth is humanity.

We maintain breeding places for disease and crime and then build hospitals and jails, organize Settlement Houses and Charitable Societies to care for those who are afflicted and those who have gone wrong.

We boast of our prosperity while 60 million of us are homeless.

We multiply our wealth and build steel walls around our treasuries to protect them from poverty and crime.

We boast of our national strength and maintain hot-houses of national dishonor, national weakness and national danger.

We do not need to educate any man to want a home. We need only to cease making it impossible for him to own one.

Give Labor Fair Play

Efficiency—When men build a factory they select a location close to raw material. They make sure that transportation will be good; that fuel will be plenty; that the market will be handy. But often they do not give a thought to the welfare of labor. And labor is the most valuable thing they use.

Ninety per cent of the value of an article is labor.

What is iron worth in a hill? It takes labor to dig it—labor to melt it and cast it—labor to make it into a plow. It takes labor to plow the ground and sow and cultivate and harvest the grain—labor to grind the grain into flour and make the flour into bread to feed labor. It is labor all the way through.

A town is not built of factories and mills, of houses and land. A town is built of men and women and these are made of boys and girls.

If we compel labor to live under crowded and unhealthful conditions, it cannot be efficient. If labor is sick, weak, indifferent, absent day after day because of illness, there is an awful loss—loss in time; loss through irregularity and inefficiency; in nursing and care; loss in doctor bills and life itself; loss through discontent and dissatisfaction; through strikes and lockouts. Human waste; human leakage—it is appalling. It means courts and jails, paupers, charity seekers, crime, disease and death.

Safeguard the Home

No company should be permitted to build a factory where it pleases. The people should decide the location. It should not be in a congested district. It should be far enough out so the workmen and their families can have a home.

The workman cannot afford to buy a home where there is no ground. He must have a place where he can grow something, have some chickens, a garden, some fruit—a place where he can make a living.

In communities where factories are not so favorably located, the city must safeguard the home and provide transportation. In these days of interurban and electric trains, of automobiles and motorcycles, the workman does not need to live close to the factory. Tenement houses adjoining mills are a disgrace, a shame. They should not be tolerated.

Pay Home-owner More

The workman who owns his home, who beautifies his ground and raises a garden is more efficient than the tenant. He is entitled to higher wages and the company should see that he gets it. The company should pay a premium on his efficiency, his steadiness, his reliability.

About 40 per cent of a man's wages is spent for food. If he has an acre of ground he will materially reduce the cost of living. Because he pays no rent and helps feed himself and his family, he is never found in the bread line and seldom moves away, if thrown out of work for a few weeks.

When a big steel mill at Pittsburgh resumed work after a brief shutdown, it was found that every workman who did not own his home, had gone elsewhere.

"The expense in breaking in new men to take their places," reports the manager, "was greater than their wages would have amounted to if they had been kept on the pay roll."

Vice and Crime—Who can paint the tragedy of unsatisfied hunger for a home? It causes despondency, breeds indifference, drives to desperation and begets defiance of law. The girl who lives in a house that is not a fit place to invite her men friends, meets them in the dance hall or on the streets. The dreary habitation has no brightness for the boy and he seeks the brilliant lights of the saloon or the gambling rooms.

Miss Harriet Fuller, superintendent of the Visiting Nurses' Association of Chicago, declares that "two-thirds of the delinquent children, two-thirds of the physically ill children, one-third of the shiftless mothers and two-thirds of the deserting fathers come from tenements with dirty and unventilated rooms. Of 50 backward children in an ungraded school, 43 resided in buildings which the State should not have allowed to exist."

The children of today are the citizens of tomorrow. How can we raise a strong and virile race; how can we stamp out crime or promote health; how can we create a great nation when we rear our boys and girls in dens of dirt and disease and degradation and death?

No place is fit to raise a boy to become an American citizen, where the mother, as she goes about her work, cannot sing "Home Sweet Home." Imagine her singing "Tenement, Sweet Tenement"?

Houses So Crowded Children Must Play in Street Thousands Killed by Autos and Cars Every Year

HOW TO PROMOTE HOME OWNERSHIP

Public Sentiment Must Be Aroused—A Square Deal and a Helping Hand are Needed

Four things are essential if we would make a community of home owners:

1. Realization.
2. Legislation.
3. Co-operation.
4. Determination.

Realization—Public sentiment in favor of home ownership must be created. We must realize the value of home ownership to the individual, to the community and to the nation.

Nothing makes a man a better husband, a better father or a better citizen than does the pride of ownership.

When he owns his home he feels independent, has more confidence, more self-respect—feels that he is an important part of the community.

His wife loves to entertain her neighbors. The children like to invite their friends.

Even though it is but a small cottage it is a real home to every member of the family. It is a place they love to be because they own it. The realization of ownership is a bond that binds them together as nothing else can, that instills within them a common interest. The father never deserts his family; the mother takes pride in being a real home-maker; the children never forget the tender associations of their home. The garden, the fruit trees, the flower beds, the cool, green lawn where they played their childish games—these always linger in their memories.

A Home Should Be Every Man's Aim

No man and wife can realize the fullness of life until they have a home of their own.

No woman should marry any man who is not willing to give her the best home he is able to give her. No man should marry any woman who will not demand that he give her such a home.

And no man should marry until he has the means to begin building or buying a home.

To buy a home may mean years of close economy and self-denial, but the bond of self-denial, the pride of ownership will draw the husband and the wife closer together than anything else can.

12

Home ownership establishes a spirit of co-operation that is good for the community. When a man owns his home he takes an interest in community development.

He believes in public improvements, in better roads, in better schools. He becomes a real citizen, respects the rights of others and upholds the laws.

He centers his interest in his home and his family; is not always moving somewhere else; sees that his home is kept in good condition; makes every sacrifice possible to educate his children.

He is contented and efficient; surrounds himself and his family with the most sanitary conditions; helps to improve the health of the community.

A city of home owners is an attractive city, a prosperous city; a healthy city; a wide-awake progressive city.

More than anything else must we realize the value of home ownership to the nation.

When the people of any country become renters, the nation goes to pieces. Tenantry killed Rome, ruined Mexico and set Russia on fire. Home ownership is the greatest factor in citizenship making: it is the cornerstone of the nation. If it crumbles, the nation crumbles.

A man will fight for his home, but he will not fight for a boarding house. He may fight in a boarding house.

The Home, The Glory of France

The Bolsheviki are sweeping Russia and Germany and Austria and Hungary. Why are they not sweeping France? Why did France stop the Germans at the Marne? Why did she stop them at Verdun? Why did she withstand the tremendous power of the greatest war machine the world ever saw? Why did France fight and fight?

Because France is a nation of home owners. Ninety per cent of the farmers of France own their homes.

A French mother sent her oldest boy to war and he was killed. She sent her second son and he was slain. She sent her third boy and he, too, was killed. She sent her fourth son and when her neighbors learned that he, also, had lost his life on the battlefield, they went to her home to comfort her. They did not find her weeping. There were no tears in her eyes, but there was a smile on her face as she kissed her last boy goodbye and sent him to battle.

"I am glad I had five sons to give to the homes of France," she said.

In time of peace, even more than in time of war we must fight for our homes. Apathy, indifference, selfishness and lack of service—these are the rocks upon which civilization may go to pieces. We must not betray our nation in its hour of greatest need.

Growth of Tenantry

America must wake up. Only 45 per cent of our people own their homes. In 1900, 46 per cent were home owners; in 1890, 48 per cent. Unless we are strong enough and patriotic enough to solve the problem, we are in danger.

The nations of the world spent $257,000,000,000 in this war. If we can raise that much to kill each other, we can raise enough to help people live.

The world will never be safe for democracy until it is a world of home owners.

Home ownership the world over is the best possible League of Nations. Only a small per cent of the people of Mexico own their homes and we have virtually been at war with Mexico for a generation. One-half the people of Canada are home owners and for over 100 years not so much as a single rifle has been needed to guard the 2,000 miles of border.

Help All to Own Homes

Legislation—It is useless to preach home ownership so long as we make it hard, almost impossible, for men to buy homes.

We made it easy for those with even the smallest salary to buy Liberty Bonds. Millions now own a share in the Government, and for the first time in their lives they feel the pride of ownership.

It is natural for men to want to own something. The babe grabs at a sunbeam because of his desire to possess it.

The promise of free land turns homeless men into frenzied mobs.

Unprincipled land owners understand this human desire for ownership, and capitalize it.

Every man craves a home of his own. If we make it easy for this craving to be realized, we will make this a nation of home-owners.

Unfair requirements, exorbitant interest—these things create discontent, endanger the safety of the nation, are criminal, unpatriotic, almost treasonable. When a man is deprived of what he has put into his home, because he was unable to meet a payment, something is wrong. The state is to blame when legalized robbery exists. Every man has a right to a square deal.

No man or corporation should be allowed to house 100 families under one roof.

Legislate Against Evils

The community that permits human beings to live under conditions that would kill a hog, is committing a crime against God and man. Such a community, such a state, such a nation invites disaster; hastens the hour of its destruction.

Every babe has the right to be born, not damned, into the world. God created the earth and the air and the sunlight for men and the children of men. And he who denies to his brother these things is defying the Almighty; cheating humanity out of its birthright.

It is the right and the duty of the people of every community to prevent the erection or use as a dwelling of any place unfit to live in.

Public sentiment against conditions that are injurious to a community must be aroused.

There must be legislation against these evils. There must be organization to see that plans approved by the community are carried out.

We Must All Co-operate

Co-operation—No movement for home ownership can succeed unless we co-operate with one another—unless we work together. Our motive must be helpfulness, not selfishness. It must be a patriotic movement. If we make it a narrow, money-making proposition, there will be no life to it. It will die—and it should die. Business must be back of it—not a part of it.

Mothers think most in the terms of boys and girls. They must make every effort, exert every influence to prevent a great and worthy idea from being exploited merely for the benefit of some real estate agent, some builder or some lumber dealer.

"Give to the world and it will come back to you." But if what we give has a string tied to it, we sink so low that the reward can never reach us.

There are real estate men and builders and lumber dealers who are inspired by genuinely worthy and patriotic motives. But there are others who are not.

It is too often the case that the man who takes the least interest in human welfare determines the policy of other men. He puts up a cheap apartment house or throws together three or four makeshift dwellings on a lot not half big enough for one family. He is the meanest of all profiteers for he speculates in human lives, prospers at the expense of human health and human happiness. His is the lowest level, but to this level he compels his competitors to sink.

We must not forget that no nation can be any greater or more enduring than its homes, and we must not allow such men as these to determine the life and stability of the nation itself.

Shall we exploit selfish interests under the guise of home building? If this is what we are undertaking, then Lord, forgive us!

Be Determined to Succeed

Determination—"Where there is no vision the people perish."

When we have the vision—when we realize that only by all working together, unselfishly, patriotically, determinedly for the common good of all, real prosperity can be ours—then and not until then will we have a community of home owners.

We must be determined—determined to awaken public sentiment for home ownership; determined to bring about conditions that will make it possible for every family to own a home; determined to see that every man gets a square deal.

If we do not own a home, we must be determined to own one. Even though our wages be small and there are many depending upon us, we can own a home if we are determined to do so.

> *Every normal man desires a home of his own. He does not merely want a roof above him and a chair beneath him; he wants an objective and visible kingdom; a fire at which he can cook what food he likes, a door he can open upon what friends he chooses. This is the normal appetite of man.—Gilbert Chesterton.*

FARM OWNERSHIP IMPORTANT

Tenantry Means Soil Robbery—Cities Depend on Farm Production

A home for every American means a home for every family residing in the country as well as a home for every family living in town. Farm tenantry is a menace to the city, the state and the nation.

The prosperity of every city depends upon the production of the farms. No farm can long be productive unless the fertility of its soil is maintained. Farm tenantry means grain farming, and grain farming robs the soil of its fertility; reduces farm production.

It takes two things to make a great country—a fertile soil and a great people. James J. Hill said that "Land without people is a wilderness, and people without land is a mob."

The Sahara Desert has no people, no homes, no schools—it is a wilderness. Russia is overrun with mobs—people without land.

When this country belonged to the Indians, the soil was fertile, but it was not worth much. Today the land is ours and we must guard and protect it. If we do not preserve its fertility we will be robbing our children and our children's children.

Farm Tenantry Ruins Soil

The farm tenant gets as much out of the land as he can. He has neither time nor inclination to build up the soil.

Every community of tenant farms shows neglect, Buildings need repair, fence corners grow up to weeds, the roads are poor. Things look like everybody was "just kind o' gettin' along somehow."

The tenant is not a community builder. He doesn't care to help in any public movement because he doesn't expect to stay long.

We must not forget that America's immense wealth is directly based upon the production of American farms.   ·

Just as sure as we become a nation of farm tenants, our prosperity, our wealth, our welfare will suffer a great loss—and the loss can never be made up.

In home ownership—farm home and city home—lies the safety of America.

17

"STAY ON THE FARM"

John Gives His Father Some Pointers on the Retired Farmer Who Moves to Town

Dear Father:

Don't move to town. Stay on the farm. You don't want to stop work. Mother doesn't want to stop work. Neither of you would be happy if you did. Be the boy and the girl on the farm. Let the children do most of the work, but you stay there.

When a farmer moves to town he's a nuisance. He's opposed to every improvement. He votes against everything—schools, libraries, paving, everything the town needs and ought to have. He kills the town.

When you hear a farmer say he is "going to move to town to educate the children," or "to give mother a chance to rest," don't you believe it. He is just moving to town to die cheap. He can stay on the farm and educate the children, and mother can get more rest on the farm in a day, than she can in town in a week. Cities don't rest people; they wear them out.

When the farmer moves to town he hasn't anything to do. He just loafs around hating himself and everybody else. He gets a grouch on—becomes a knocker. But he keeps on eating—doesn't get enough exercise to help digest his food, and begins to feel miserable. Then it isn't long until he is "all in." He has a funeral—a hearse and one carriage.

If he had died on the farm, where he had always lived, the whole township would have turned out. Everybody would have mourned a good man gone. But the town won't miss him.

Meanwhile, he has rented the farm, and the old place is going to pieces. It has become a grain farm. The soil is robbed—the buildings need paint, the fences are falling down, the barn-door hangs on one hinge.

It's a scrub farm. Down the road is a scrub school-house, with a scrub teacher.

No, Father, you stay on the farm. Don't kill the farm by moving off of it, and kill the town by moving into it.

Your loving son,

JOHN.

EDUCATE PUBLIC OPINION

What is the glory of our architecture, if the poor must hide in dens and holes? What is our boast of greatness and strength, if the weakest are not cared for? What is our pride in mental achievement, if the thought of the people tolerates filth and degradation? What is our advantage in wealth if poverty and crime threaten our treasuries? Let us wipe out the shame of our cities, and take away the reproach of the poor. Let us make this a nation, not of tenements, but of homes.

—Albion Fellows Bacon.

With Americanism as our guide, with red blood in our veins, with Christian faith in our souls, let us go forth today and do our duty. Let us attract the attention of those who have been too busy to think. Let us convince those who are negligent and skeptical. Let us plead with those who have been too busy. Let us fight those who would profit from the misfortunes of our brother. Some of us have good intentions but very little determination. Advancement can come only through educated public opinion.

Too much "good time"

Too much indulgence

Too much idleness

Too much ease

Too much "do as you please"

Will ruin any boy.

We build jails when we should be building schools of the right kind. We try to reform the boy instead of educating him.

We must build children as well as homes.

The tenant doesn't take root. He is always asking himself, "Where do we go from here?"

19

The workman that invests his money in a real estate deed instead of in rent receipts is a real asset to the community.

Don't waste time fighting weeds or evil things. Just plant something worth while and take care of it. It is direction, not correction, that is needed.

The way to reform the boy is to encourage him to do right. Give him a real job. Tell him you picked him out to do it because it is a hard job. He'll make good.

No man can gain real success at the expense of others.

Cities try to economize by crowding hundreds of families into one acre—and then spend billions for parks and playgrounds.

"It's none of our business how the other half lives." It is decidedly the business of every one of us.

What you do may decide the future of other states and many cities.

Are you afraid to do your duty?

"Am I my brother's keeper?"

Too often selfishness and greed have filled the pockets of the owner and wronged the tenant until he has lost faith in his brother man.

What Must Be the Effect on a Child's Character as Well as His Health to Be Brought Up Amid Surroundings Like These?

THE REAL HOME

By Edgar W. Cooley

Somewhere, before the dawn of civilization, a man and a woman sought a safe place for a little child. It may have been a cave, a crevice amid the rocks, a tangle of grass in a jungle, or a dark recess in the forest. It may have been furnished only with matted leaves or the skins of animals, but it was a place where the family could find comfort and companionship; a refuge in which the children could grow and develop—the one spot to which they laid absolute claim of ownership, over which the mother watched with tender care; for which the father fought and would have died. It was the world's first home.

During the millions of years since then, worlds have been destroyed, continents have disappeared, nations have been overthrown. But the home has survived.

Many millions of men, women and children inhabit the earth The children outnumber the adults by the ratio of three to two, and the home and the school shape the destiny of boys and girls.

The real home is not simply a place to stay. It is a place the children will always remember; a place to which they will love to return, in which they will rejoice to live. Because it belongs to and is a part of the family, it is a real home.

The rented dwelling, the apartment house, the flat, can never be a true home. It can never center within itself the affections of a generation. It can never impart that stimulus which the pride of ownership inspires. More serious than all else, it cannot cultivate, to its fullest maturity, the love of the parents for their children.

The home is the institution for which and by which all other institutions exist. It was the birthplace of liberty. It is the shrine of patriotism and the abiding place of love and peace and true and lasting friendship.

WHAT THE WOMEN CAN DO

ORGANIZE
GIVE PERSONAL SERVICE
HAVE A DEFINITE PLAN
HAVE A WAYS AND MEANS COMMITTEE
HAVE A PUBLICITY COMMITTEE
HAVE A BUILDING COMMITTEE
PUT ON A STATE-WIDE CAMPAIGN
SECURE THE GOVERNORS PROCLAMATION
CONDUCT PRIZE-ESSAY CONTESTS

A HOME IS THE BIRTHRIGHT
OF EVERY CHILD

MOTHER IS THE HOME MAKER

OTHER THINGS WE CAN DO

VITALIZE RURAL SCHOOL WORK

KNOW WHAT
      THE BOYS AND GIRLS ARE DOING

LET THEM HELP PLAN THE WORK

ENCOURAGE OWNERSHIP

ESTABLISH COMMUNITY LAUNDRY

PUT ON A FLY CAMPAIGN

FURNISH PLANS FOR SEPTIC TANKS

INSTALL REST ROOMS

BEAUTIFY TOWN AND COUNTRY

CLEAN UP PAINT UP FIX UP

DEMONSTRATE HOME CANNING

THE FOLLOWING IS FOR HISTORICAL PURPOSES ONLY
AND DOES NOT REFLECT CURRENT
SCIENTIFIC KNOWLEDGE, POLICIES,
PRACTICES, METHODS OR USES.

HELPS
FOR
WASH
DAY

Headwork

Lightens

Housework

PUBLISHED 1916 BY
INTERNATIONAL HARVESTER COMPANY
OF NEW JERSEY (INCORPORATED)
AGRICULTURAL EXTENSION DEPARTMENT
HARVESTER BLDG., CHICAGO

FOREWORD

There are thousands of American homes which still observe the time-worn, slavish custom of "Wash Day." Of course, there are few men to-day so neglectful of their wives as to not provide the few inexpensive things necessary to make their housework easier.

Any man can buy his wife a washing machine run by kerosene power. This will do away with the old zinc-top washboard run by elbow grease. Are you using a hoe to do your cultivating and a crooked stick for plowing—or have you a two-row cultivator, and a sulky plow?

Since we are discussing "Wash Day," we might say that while a washing machine is better than a washboard, it would be better still if you got busy with the neighbors and organized a Community Laundry. Others have done it. You can do it.

There are numerous devices which a man can bring on his trips from town; and it will not require any high financing to get them, either.

Have you installed a lighting system and consigned the kerosene lamp to the junk pile, or are you going to let your wife go on cleaning smoky lamp chimneys for the rest of her life?

Have you piped the water in the house? It's a safe bet that you have water piped in the barn for the stock. Why not in the house; your wife uses water?

How about that bathroom, cement cellar, wash-house, kitchen cabinet, bread-mixer, fireless cooker, hot water heater, vacuum cleaner, electric fan, electric iron, gasoline range and cream separator? How about that new churn and kerosene engine to run it? All of these things should be in every rural home.

Think these things over. But do not stop there. Get busy. You will be a better citizen in the community and you'll certainly feel better for it.

HELPS FOR WASH DAY

Things We Can Do to Take the Drudgery Out of Wash Day—Headwork Lightens Housework

There are two objects in view in laundering clothes. The first is cleanliness, the second is improving the appearance of the garments. In the first, soap and water are the only essentials, in the second, enter starching and ironing. Clothes of cotton, linen and silk are easily laundered and for this reason are used next to the body, as frequent changes are desirable.

The washing of clothes may be a simple or complex work, depending upon the desires of the family, the skill of the laundress and the equipment of the laundry. It can be done in a wash stand, in a wash basin, with movable tubs, or permanent fixtures, so the process of laundering naturally depends upon the kind of appliances with which it is to be done.

Washing is really a most flexible labor. It can be done with many appliances or it may be done without any. Rubbing clothes by hand in the primitive way, or, sitting on a stone on the bank of the river, one can wash clothes and wash them clean, provided enough water passes through the fiber of the clothes. Soap hastens the cleansing, warm water helps it, the washboard facilitates the rubbing, and the washing machine increases the speed of the cleansing process.

Preparing to Wash

Before wash day, clothing should be thoroughly inspected to discover rents and stains, carefully sorted, and the white clothes put to soak.

The following outline is suggested for the preparation of clothes for washing:

1. Sort the clothes according to kind:
    (a) White cotton and linen clothing.
         Table linen and slightly soiled towels.
         Bed and body linen.

3

Handkerchiefs.
Soiled towels and cloths.
(b)   Colored clothing.
(c)   Flannels.

2. Mend rents, except in stockings.

3. Remove stains.

4. Put as many white clothes to soak as is practicable. Some colored clothes having fast colors may be soaked if very much soiled.

The purpose of soaking soiled clothes before washing them is to soften and separate the fibers of cloth in order to loosen the dirt. Water alone accomplishes the purpose to a great extent; but the use of soap, or a soap solution to which has been added borax, ammonia, or other alkali, and turpentine, kerosene or benzine, makes the washing process both easier and quicker.

How to Save Time

It is well before beginning the washing to make a soap solution, as it gives a quick suds and is more easily handled, and its use will therefore save time.

All the clothing should not be put to soak in one tub. If three tubs are available soak the table linen and slightly soiled towels in one, bed linen and body linen in second, soiled towels and cloths in third. If only two tubs are available, wash table linen and slightly soiled towels without preliminary soaking. Soiled towels and cloths should always be soaked before washing.

If members of the family have colds, the handkerchiefs should be put to soak in a solution of boric acid in a basin by themselves, and should be washed separately and boiled for twenty minutes.

Wet the garment to be soaked, rub the more soiled part with soap solution, and fold that part in. Fold and roll each garment separately and pack it into the tub with the other garments. Folding and rolling prevents the dirt in the soiled parts from spreading. Cover the clothes with warm, soapy water, to which may have been added an alkali such as borax or ammonia, and an oily substance, perhaps turpentine, kerosene, or benzine. Directions for making soap solutions are given under the heading, "Soap and Soap Solutions" (p. 7). Cover the tub, and if possible let the clothing soak in it several hours or over night. If colored clothes are to be soaked, cover with warm water or with water slightly soapy. No alkali should be used with the colored clothing.

No arbitrary order can be recommended for washing clothes, but flannels, white goods, and colored goods, should be washed separately, as the washing process differs somewhat for each case.

Why to Maintain an Even Temperature

A few simple explanations may aid the housekeeper in solving some. of her problems. Heat tends to expand the threads of the cloth, and the expansion aids in removing the dirt caught between the threads. If the cloth is cooled during the washing process, the thread contracts and the dirt is again entangled; consequently, after the cloth has once been warmed, one of the objects of the launderer should be to maintain an even or a rising temperature. In the commercial laundry an even temperature is kept by turning the right amount of steam into the washing machine. In the home laundry, boiling water added from time to time will aid in keeping an even temperature. A good suds is necessary in the washing process. As the suds falls, that is, as it is used up by uniting with dirt, more suds should be supplied by adding more soap or soap solution. If sufficient soap is not used, insoluble black specks are often left on the clothing.

All utensils, receptacles, and apparatus should be immaculately clean.

Plenty of Hot Water is Essential

Have plenty of hot water before beginning the washing. If possible, soft water should be used; if water is hard, soften it as directed.

1. For each gallon of water, use two tablespoons of a solution made by dissolving one pound of washing soda in one quart of boiling water. The solution may be bottled and kept on hand, and is a useful cleansing agent.

2. For each gallon of water use one-fourth pound of lye dissolved in one cup of water.

3. For each gallon of water use one teaspoon of borax dissolved in one cup of water. If water is very hard increase the amounts used.

Outline for Washing White Linen and Cotton Clothes

1. Put water on to heat.

2. Make soap solution (use one cake of soap to two or three quarts of water).

3. Rinse clothes from water in which they have been soaked.

4. Wash clothes from warm suds in the following order:
   (a) Table linen and slightly soiled towels.
   (b) Bed linen.

(c)   Body linen.
(d)   Handkerchiefs.
(e)   Soiled towels and cloths.
(f)   Stockings (white).

5.   Wash again in clean suds. Wring.
6.   Boil in clean, slightly soapy water.
7.   Rinse in clean, clear water. Wring.
8.   Rinse in bluing water. Wring.
9.   Starch.
10.   Hang to dry.
11.   Remove from line, dampen and fold.

Whenever the water becomes dirty, use fresh suds. Clothes cannot be made clean without the use of plenty of water. Keep up a good suds while washing, and add hot water from time to time. If a washing machine is used, do not put enough water in the machine to float the clothes; if you do, they escape the mechanical action of the dasher and are not sufficiently rubbed. Clothes should be wrung from the wash water through the wringer. The screws of the wringer should be adjusted to bring its rolls close together and clothing should be folded so as to give it an even thickness in passing through the wringer, for heavier garments loosen the screws of the wringer. Fold in buttons and hooks and turn the wringer slowly.

For White Wool and Flannel

White woolen garments, serges and flannels should be cleaned with one pound of white soap and three ounces of powdered borax added to five gallons of water. Put them in two solutions of this and two warm rinsing waters, and then through a third rinsing solution consisting of one-third of a pint of acetic acid, one-half ounce of oxalic acid crystals and ten gallons of water. If still yellow, soak for a few hours in a bath of peroxide of hydrogen (one part to twenty parts of water). Rinse, wring between two cloths and hang in the open air but not in the sun.

Lace Curtains

To clean lace curtains wash them carefully in a solution of one pound of soap, two ounces of sal soda and nine gallons of water. Place them for ten minutes in a boiling solution of white soap. Rinse well, and if necessary, let the curtains lie until they are white in the following liquid: Dissolve one-half pound of chloride of lime and one-half pound of sal soda in two gallons of water, boil this ten minutes, strain, and add ten gallons of water. Put them through a weak hot starch (one ounce to two gallons of water), adding a small piece of wax—a bit of candle is excellent. Wring, place on frame and press the edges with an iron.

Blankets

Blankets cleaned in the following way will be as soft as new: Put them in two warm solutions of white soap, using one pound of soap and two ounces of borax to each eight gallons of water. Follow with two warm rinsings and a warm solution of one-half ounce of oxalic acid and one-third pint of acetic acid to fifteen gallons of water. If the blankets are colored, omit the oxalic acid, otherwise they are cleaned just as white ones. Dry in the open air if possible. To make them look particularly well when they are dry, lay them on a table and brush the nap in one direction on both sides.

Soap and Soap Solutions

Rubbing the bar soap on the garment is rather an expensive if not extravagant method of applying the soap. The better if not the safer plan is to prepare a soap solution, because of economy, and also because the contact of soap often discolors fine fabrics.

For Washing Colored Goods

One-quarter pound mild or medium soap to 1 gallon water.

For Ordinary Purposes

One bar ordinary washing soap, 2 or 3 quarts of water. Shave soap and put into saucepan with cold water. Heat gradually until soap is dissolved (about 1 hour).

For Soaking Clothes

One bar ordinary soap, 3 gallons water, ½ to 1 tablespoon turpentine, 1 to 3 tablespoons ammonia.

For Washing Much-Soiled Woolens and Delicate Colors

One-half pound very mild or neutral soap; ¼ pound borax, 3 quarts water.

Soap Jelly. With Turpentine Incorporated

One bar soap, 1 quart water, 1 teaspoon turpentine or kerosene.

Soap Bark

One pound of soap bark equals 2 pounds of soft soap and is excellent to use in place of soap.

Bran

One quart water, 1 cup bran. Boil ½ hour. Strain, when needed reduce with warm water.

Home-Made Soap

One pound lye dissolved in 3 pints of cold water, 5 pounds fat melted, 1½ tablespoons borax, ½ cup ammonia. When lye mixture has cooked, add it to fat, stir until the mixture is as thick as honey, pour into wooden or pasteboard boxes lined with oiled or waxed paper, set away to harden.

Javelle Water.

Javelle water forms a very efficient bleaching liquid for unbleached fabrics, as well as for cotton goods that have become yellow with dirt and age. It is made as follows:

1 lb. washing soda.	½ lb. chlorid of lime.
1 qt. boiling water.	2 qts. cold water.

Put the soda in an agate—never an aluminum—pan, and add the boiling water. Mix the lime in the cold water. Let the mixture settle and pour the clear liquid into the dissolved soda. Bottle and keep in a dark place.

Removing Stains

Stains should be removed from the clothes before the soap is applied, if possible. The process of removing stains is fundamentally the same as that of removing other forms of dirt; that is, to find some substance in which the stain is soluble or which will aid in its removal. Following is a list of solvents which are valuable in removing stains which resist ordinary washing:

Salt	Milk
Ether	Borax
Benzol	Vinegar
Alcohol	Carbona
Kerosene	Sunshine
Ammonia	Oxalic acid
Turpentine	Chloroform
Lemon juice	Javelle Water
Naphtha soaps	Ink eradicator
Olive oil, lard, etc.	Hydrogen peroxide
Water, both hot and cold	Hydrochloric acid (a strong
Benzine, naphtha or gasoline	acid, very corrosive to fab-
Fuller's earth and French chalk	rics and flesh)

Methods of Removing Stains

Blood

1. Wash in cold water until stain turns brown, then rub with naphtha soap and soak in warm water.

2. Rub with common soap, then soak in cold water to which a teaspoon of turpentine has been added.

3. If the goods is thick, apply a paste of raw starch to the stain. Renew paste from time to time until stain disappears.

Chocolate

Sprinkle with borax and soak in cold water.

Coffee

Spread stained surface of the cloth over bowl or tub. Pour boiling water through the stained part of the cloth. Pour the water from a height so as to strike the stain with force.

Cream

Wash in cold water, then with soap and water.

Fruit and Wine Stains

1. Treat with boiling water as for coffee.

2. If the stain resists the boiling water treatment, soak the stained part of the cloth for a few minutes in a solution made from equal parts of javelle water and boiling water. Rinse thoroughly with boiling water to which a little dilute ammonia water has been added. Repeat if necessary.

Grass Stains

1. Soak in alcohol.

2. Wash with naphtha soap and warm water.

3. If the fabric has no delicate colors and the stain is fresh, treat with ammonia water.

4. For colored fabrics, apply molasses or a paste of soap and cooking soda. Let stand over night.

Grease Spots

1. Wash thoroughly with naphtha soap and water.

2. Soften old grease spots with turpentine, oil, or lard before washing the cloth.

3. Dissolve the grease in benzine, alcohol, chloroform, ether, carbona, or benzol.

4. For delicate fabrics dissolve grease spots in ether or chloroform. Chloroform and carbona are useful because noninflammable.

5. Apply a paste of fuller's earth or chalk to absorb grease.

Indigo
Treat as for coffee.

Ink
Ink is often difficult to remove, as it varies greatly in composition. It is well to experiment with a corner of the spot before operating on the whole.

1. If the ink stain is fresh, soak the stained portion of the cloth in milk. Use fresh milk, as the old becomes discolored.

2. Wet the stain with cold water. Apply a ten per cent solution of oxalic acid to stain, let stand a few minutes, and rinse. Repeat until stain disappears. Rinse in water to which borax or ammonia has been added. (Oxalic acid is a very poisonous substance.)

3. Javelle water will remove some ink stains. Apply as for rust stains.

4. Treat with hydrochloric acid as for iron rust.

5. Treat with lemon juice and salt, as for iron rust.

6. Use alcohol for some ink stains.

Milk is the only agent given that does not remove color.

Iodine Stains
Soak in alcohol, chloroform, or ether.

Iron Rust
1. Wet the stained part with borax and water, or ammonia, and spread over a bowl of boiling water. Apply a ten per cent solution of hydrochloric acid, drop by drop, until the stain begins to brighten. Dip at once into alkaline water. If the stain does not disappear add more acid and rinse again. After the stain is removed, rinse at once thoroughly in water to which borax or ammonia has been added. The borax or ammonia is to neutralize any acid that may linger. Less dilute acid may be used if the operator is skillful.

2. Proceed as with hydrochloric acid, but use a ten per cent solution of oxalic acid instead of hydrochloric acid. Oxalic acid is not so detrimental to fabrics as is hydrochloric acid, but it is a deadly poison even in dilute solution.

3. Wet the stained part with a paste made of lemon juice, salt, starch, and soap, and expose it to sunlight. This is a simple method to employ, but it takes longer and is often not effective.

4. Soak stain in javelle water for a few minutes, then wash. Repeat until stain disappears. Javelle water is weaker in action than is hydrochloric acid. All the iron-rust-removing substances destroy color, and unless care is taken will greatly weaken the fabric.

Lampblack

Saturate spot with kerosene. Wash with naphtha soap and water.

Machine Oil

1. Wash with soap and cold water.

2. If the stain does not respond to the soap-and-water treatment, use turpentine as directed for paint stains.

Meat Juice

Wash in cold water, then with soap and water.

Medicine Stains

Soak in alcohol.

Mildew

Mildew is very difficult to remove if of long standing.

1. Wet stains with lemon juice and expose to sun.

2. Wet with paste made of one tablespoon of starch, juice of one lemon, soft soap, and salt, and expose to action of sun.

3. Treat with paste made of powdered chalk and expose to action of sun.

Milk

Treat as directed under cream.

Mucus

Soak in ammonia water or in salt and water, then wash with soap and cold water.

Paint

1. Wet the spot with turpentine, benzine, or alcohol, let it stand a few minutes. Wet again and sponge or pat with a clean cloth. Continue until stain disappears.

2. For delicate colors treat with chloroform.

3. If the paint is old it may take some time to soften. Treat old paint stains with equal parts of ammonia and turpentine.

Perspiration

1. Wash in soapsuds and expose to the action of sunshine.

2. Treat with javelle water as directed for iron rust.

3. Treat with oxalic acid as directed for iron rust.

Scorch

Scorched fabrics can be restored if the threads are uninjured.

1. Wet the stained portion and expose to the action of the sun. Repeat several times.

2. Extract juice of two onions, add one cup of vinegar, two ounces fuller's earth, and half an ounce soap. Boil. Spread paste over scorched surface. Let it dry in sun. Wash out thoroughly.

Stove Polish
1. If fresh, remove by washing.
2. If the stain is old, treat as directed for tar and lampblack.

Tar
Treat as directed for lampblack.

Tea
1. Treat as directed for chocolate.
2. Soak the stain in glycerine, then wash.

Varnish
Treat as directed for paint.

Vaseline
Wash with turpentine. Boiling sets this stain.

Wagon Grease
Soften with lard or oil and wash in soap and water.

To Set the Color in Cotton Material

To set the color in cotton materials let the goods stand for twenty minutes in a solution made in the following way: one pound of sugar of lead (be careful, this is poison) in a gallon of cold water, and one pound of alum in a gallon of cold water. Combine and add one pint of this liquor to every three gallons of cold water.

Bluing

Bluing is used to counteract the tendency to yellowness of white clothes. The amount used in any case varies according to the material and weave.

The most common forms of bluing are:
Indigo.
Ultramarine.
Prussian.
Aniline.

Indigo blue is almost impossible to purchase in the market. Ultramarine blue comes in ball or block form and is the most satisfactory of all blues.

Prussian blue usually appears in liquid form, only occasionally as a powder. The color is greenish blue. This bluing is a favorite with many a family laundress. If used after clothes have been carelessly rinsed, the iron in its composition will probably prove troublesome, as the soap which is carried into the bluing water makes an iron compound, which later appears as rust spots on the clothes. When compared with other bluings, this is by no means as cheap a bluing as many consider it.

To Use

Before using, the bluing should be stirred each time to blend. In case of the indigo and ultramarine blues, settling of particles of blue is very noticeable even if bluing has stood for only a short time. Each article washed should be shaken out before putting into the blue tub. Only a few pieces should be put in at a time. Stirring the blue each time it is used, shaking out each piece, and leaving in the blue for a few minutes only, will prevent streaking. It is better to dip a garment several times rather than allow it to stay in the blue tub for any length of time.

Effect of Bluing on Different Weaves

Materials that are open in weave, as table linen, laces, etc., will take the blue very readily; therefore, bluing for them should be light in tone. For closely woven material, as sheets, etc., the blue should be considerably deeper.

To Remove Blue

Clothes that are too blue, as a result of bluing being made too deep in color or from an accumulation of repeated bluings, should be put into boiling water and allowed to remain for a half hour. When the excess blue will not yield to this treatment, clothes should be boiled.

Tinting

For very dark blue or black material, the bluing should be made very deep in color in order to be of any use to these colors.

To Restore Flesh Colored Crepe de Chine Waists

Put a piece of pink crepe paper in the water. Test the color with a piece of cheesecloth and when of right shade immerse the waist in the colored water.

To keep white crepe de chine waists from turning yellow after washing, wrap them in a Turkish towel over night and in the morning they are just moist enough to iron. Hanging in the air is what makes them yellow, and if done this way they will keep the dead white of new material.

Starching

Starch is in the form of minute compact granules, insoluble in water, obtained from many plant tissues. We are familiar with the powder that a mass of these granules form. When heat causes the moisture to penetrate the granules, they swell, burst, and form a thick, sticky mass known as starch paste. Starch has the power of penetrating the pores of the fabric. The kind of starch used determines its penetrative power. On drying, it gives to clothing a characteristic stiffness.

Starch is made from a number of materials and there has been more or less discussion concerning that which is best.

Starches are rated according to their ability to make a paste which penetrates the fiber of the fabric and at the same time resists moisture. This is called viscosity.

Starch is made from corn, wheat, rice, potatoes, etc., etc.

The purpose of the launderer is to blend starch with the fabric in such a way as to make the starch a natural part of the cloth; to give the desired degree of stiffness and yet keep the fabric pliable; to give a body as enduring as possible and capable of resisting moisture; to give clearness, good color, and any desired finish, whether dull or glazed.

Various substances are used with starch to increase its penetrability and prevent it from sticking to the iron, as well as to give pliability to the cloth, increase its body, and improve its color. Of these substances may be mentioned borax, alum, paraffin, wax, turpentine, kerosene, gum arabic, glue and dextrin.

Borax in Starch

Borax increases the penetrability of starch and prevents its sticking to the iron. Moreover, starch containing borax adds gloss to a garment, increases its whiteness, and gives it greater body, and more lasting stiffness, than it would otherwise have.

Alum

Alum is used alone, or with borax, in starch to improve color, to increase penetrability and pliability, and, last but not least, to thin the starch mixture. When alum is cooked with a starch paste it causes the paste to become thinner. ''Cooking thin'' with alum does not affect the strength of the starch mixture and is an advantage when a stiff starch is desirable and the thick mixture would be inconvenient to handle. By the use of alum, starch may be made thin without dilution. Alum has been objected to by some persons as being somewhat injurious to fabrics.

Wax, Paraffin, Turpentine, Lard, Butter

Oily substances are used to add a smoothness, gloss and finish; to prevent the starch from sticking to the iron, and to aid in preventing the absorption of moisture.

A Candle in the Starch

To secure smoothness and glossiness when ironing starched pieces, stir the starch three or four times, while boiling and just ready to remove, with a paraffin candle.

Gum Arabic, Glue, and Dextrin

Substances resembling glue are used with starch to increase its stiffening power. They are sometimes used alone when the white color of starch is considered a disadvantage in stiffening colored fabrics.

Directions for Using Starch

In making starch a naturally soft water is greatly to be desired, but if the water is hard it should be softened by borax, not with washing soda or lye, since soda and lye tend to produce a yellow color with starch.

1. One-fourth cup starch to 1 quart water gives moderate body stiffness.

2. One-half cup starch to 1 quart water gives stiff body finish.

Directions for Cooking Starch

Starch should first be mixed with a little cold water until thoroughly dissolved and then stirred slowly into boiling water and cooked 15 or 20 minutes.

Thorough cooking of starch is very desirable in laundry practice, for it increases the penetrability of the starch and decreases its tendency to stick to the iron. If borax, lard, kerosene, or other like substance is used, it should be cooked with the starch, to insure thorough mixing.

Thick starch:

> One-half cup starch, mixed with ½ cup cold water.
> One quart boiling water.
> One-half to 1 level teaspoon borax.
> One-fourth level tablespoonful lard or butter or kerosene or turpentine; or ¼ inch square wax or paraffin.
> Mix, and cook as directed under directions for cooking starch.

Thin starch:

> One-half cup starch, mixed with ½ cup cold water.
> Three quarts boiling water.
> Other ingredients, same as for thick starch.
> Mix, cook as directed under directions for cooking starch.

Clear starch:

> Dilute ½ cup thick starch with 1 quart hot water.
> Clear starch is used for thin muslins, infants' dresses, etc.

Cold starch:
Same proportions as for thick starch.
Use borax but omit fatty substances.
Stir thoroughly before using.

Raw or cold starch is often used with very thick or very thin goods, to increase their stiffness. A fabric will take up a greater amount of starch in the raw form than in the cooked form. The desired stiffness is produced by the cooking given the raw starch by the heat of the iron. The difficulty of ironing is increased by using raw starch, for unless the ironer is skilled, the starch cooks on the iron and starch specks are then produced on the clothes.

Wringing and Drying

The wringing of clothes is also important and a good wringer is an essential part of good laundering. The rolls should be of the best rubber possible, mounted on ball bearings so as to operate easily. A good wringer squeezes the water from the fabric and so does not injure the fabric as does a wringer that twists the fabric, or the hand that actually wrings the water from the fabric.

Now to the drying, in which a good clothes line is absolutely essential. Choose one of cotton that won't twist or break. Get the best one you can buy and cherish it carefully.

Sponging, Pressing and Removing Spots

Dark-colored suits, when not badly soiled, can be sponged and pressed. To do this brush them well with a good stiff brush to remove all loose dust and dirt. Then sponge thoroughly till they are well moistened, with a lukewarm mixture of one part of household ammonia to three parts of water; brush well with a good brush, hang up to dry, and finally press with a heavy iron, laying a cloth over the goods. Any grease spots not yielding to this process may be removed by using gasoline.

To sponge new cloth, take a heavy cotton cloth, wet it thoroughly and wring it out. Spread your goods on a table folded into four thicknesses, and lay the damp cloth over it, ironing with a heavy iron until one side is done; repeat on the other side.

Four excellent remedies to have ready are gasolene, strong ammonia, acetic acid and oxalic acid. These acids are poison and should be carefully labeled and put in a safe place. They will keep for a long period.

When removing a spot do not begin to work right on the spot, but start some distance away and gradually work toward it, using a good brush. In this way you will avoid leaving a ring. If in spite of this precaution a ring is left sponge the whole thing in gasolene.

Dark-colored garments which have had the color entirely removed in spots by acids may be restored by an application of ammonia, and to those which have been spotted by alkalies, an application of diluted acetic acid is beneficial. A garment spotted by the rain may be freshened by laying the garment over an ironing board and steaming it. This is done by placing a damp cloth over the article and ironing.

To remove blood stains, or glue stains on woolen, cotton, linen or silk, wash with warm water to which one teaspoonful of soda to each quart of water has been added.

To remove varnish or paint on wool, cotton or linen, rub carefully with gasolene and soap; on silk use gasolene.

To Clean Ribbons

Satin, silk brocade and ribbons should be cleaned by the dry cleaning method. White silk or satin only slightly soiled may be cleaned by brushing with bread crumbs, powdered starch or magnesia, and then dusting well. White chiffon, silk crepe and similar materials should be dry cleaned.

Cleaning Gloves

To clean kid gloves lay them on a board or draw them on the hand and brush with the soap solution and then with clear gasolene. After this pass them through a wringer between two clean cloths, pull them into shape and hang in the open air to dry. After drying dust white gloves with powdered pipe clay, chalk or magnesia.